A First Course in Network Science

Networks are present in all aspects of our lives: networks of friends, communication and transportation networks, and the Web are all examples that we experience outwardly, while the neurons in our brain and the proteins within our body form networks that determine our intelligence and survival. This modern and accessible textbook introduces the basics of network science required across a wide range of job sectors from management to marketing, from biology to engineering, and from neuroscience to the social sciences. Students will develop important, practical skills and the ability to write code for using networks in their areas of interest — even as they are just learning to program with Python. Extensive sets of tutorials and homework problems provide plenty of hands-on practice and longer programming tutorials online further enhance students' programming skills. This intuitive and direct approach makes the book ideal for a first course, aimed at a wide audience without a strong background in mathematics or computing but with a desire to learn the fundamentals and applications of network science.

Filippo Menczer is Professor of Informatics and Computing at Indiana University, Bloomington. He is an ACM Distinguished Scientist and board member of the Indiana University Network Science Institute (IUNI). He serves in editorial roles for the leading journals *Network Science*, *EPJ Data Science*, and *PeerJ Computer Science*. His research focuses on network science, computational social science, and Web science, with a focus on countering social media manipulation. His work on the spread of misinformation has received worldwide news coverage.

Santo Fortunato is Director of the Network Science Institute (IUNI) and Professor of Informatics at Indiana University, Bloomington. His current research is focused on network science, specifically network community detection, computational social science, and the 'science of science'. He received the German Physical Society's Young Scientist Award for Sociophysics and Econophysics in 2011 for his important contributions to the physics of social systems. He is Founding Chair of the International Conference on Computational Social Science (IC2S2).

Clayton A. Davis holds a Ph.D. in Informatics and BS and MA degrees in Mathematics from Indiana University, Bloomington. His research is concerned with the development of big-data platforms for social media analytics, machine learning algorithms for combating online abuse, design of crowdsourcing platforms, and the role of social media in social movements. His work on social bot detection was featured in major news outlets worldwide. His Web tools, including Botometer, Kinsey Reporter, and the Observatory on Social Media, answer millions of queries from thousands of users weekly. He won the 2017 Informatics Associate Instructor Award for his role in the development of high-quality teaching material for network science courses.

A First Course in Network Science

FILIPPO MENCZER

Indiana University, Bloomington

SANTO FORTUNATO

Indiana University, Bloomington

CLAYTON A. DAVIS

Indiana University, Bloomington

CAMBRIDGE
UNIVERSITY PRESS

CAMBRIDGE
UNIVERSITY PRESS

University Printing House, Cambridge CB2 8BS, United Kingdom

One Liberty Plaza, 20th Floor, New York, NY 10006, USA

477 Williamstown Road, Port Melbourne, VIC 3207, Australia

314–321, 3rd Floor, Plot 3, Splendor Forum, Jasola District Centre, New Delhi – 110025, India

79 Anson Road, #06–04/06, Singapore 079906

Cambridge University Press is part of the University of Cambridge.

It furthers the University's mission by disseminating knowledge in the pursuit of education, learning, and research at the highest international levels of excellence.

www.cambridge.org
Information on this title: www.cambridge.org/9781108471138
DOI: 10.1017/9781108653947

First published 2020 (version 2, January 2023)

Printed in Great Britain by Ashford Colour Press Ltd.

A catalogue record for this publication is available from the British Library.

ISBN 978-1-108-47113-8 Hardback

Additional resources for this publication at www.cambridge.org/menczer.

Colleen, Massimiliano, Iris: thank you.
—*Filippo Menczer*

To my parents and my brother.
—*Santo Fortunato*

Liz, Gina, Mary Jo and Jay: your love
and support mean everything to me.
—*Clayton Davis*

Contents

Preface

Networks are present in all aspects of our lives: networks of friends, communications, computers, the Web, and transportation are examples that we experience outwardly, while our brain cells and the proteins within our body form networks that determine our survival and intelligence. When people communicate through Facebook or Twitter, buy stuff on Amazon, search on Google, or buy an air ticket to visit family, they use networks without knowing it. Today, a basic understanding of network processes is required in job sectors from technology to marketing, from management to design, and from biology to the arts and humanities. This textbook explores the study of networks and how they help us understand the complex patterns of relationships that shape our lives.

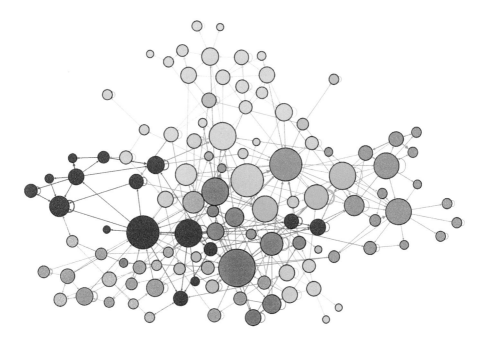

This book is also a network! The relationships between chapters, sections, and subsections are depicted in the above image. Links represent both the hierarchical structure of the book (as seen in the Contents) and cross-references among chapters, sections, figures, tables, equations, and boxes. Node colors represent chapters and node size is proportional to the number of neighbors.

Why a "First Course" in Network Science?

This is not the first book on network science — in fact there are several excellent ones to choose from, and we list a few in Chapter 1. We have been teaching these topics for several years at Indiana University, to a broad audience of undergraduate students in informatics, computer science, data science, information science, business, and the natural and social sciences. This experience has taught us that students are eager to "get their hands dirty" and write code to both understand and use networks in their application domains of interest — even as they are just learning to program and lack math and computing background beyond high school and entry-level college courses. So we developed a wide set of tutorials and problems, both theoretical and computational, providing students with an abundance of hands-on practice in network science. Using such an approach, the book introduces networks to a wide audience of learners with no technical prerequisites other than some introductory programming, and a willingness to learn by doing. This makes the textbook appropriate for a "first course" in network science.

Synopsis

After surveying networks in many areas, we talk about social networks, which are the most familiar to students. This allows the introduction of concepts like the small-world property (short paths) and clustering (triangles and transitivity). These topics are explained through fun learning activities such as the *Six Degrees of Kevin Bacon* game. Then we delve into the role of hubs by using the Friendship Paradox and discuss network robustness. Next, we introduce directed and weighted networks, respectively. The Web, Wikipedia, citations, traffic, and Twitter are used to illustrate the role of direction and weights. The final three chapters cover more advanced topics, namely models for the emergence of networks, community detection methods, and dynamic processes that take place on networks.

Each chapter brings into focus the basic concepts necessary to understand a fundamental aspect of networks; advanced topics and formalism are avoided. When a bit of math is helpful, we include it in boxes. This slightly more technical content can be skipped without loss of basic understanding of a topic. But students who can follow these extra notes will be able to gain a deeper comprehension of the material. Each chapter includes programming tutorials and exercises, allowing readers to apply and test their knowledge through hands-on activities for building and analyzing networks. These tutorials work on examples of real-world networks that are used to illustrate concepts throughout the book. Both tutorial code and network data are available on the book's GitHub repository.[1]

---

[1] github.com/CambridgeUniversityPress/FirstCourseNetworkScience

Target Audience

With the explosion of popularity and commercial success of online social media, many students are interested in learning a bit about what is "under the hood" of networks. This textbook is aimed at all of those students, mainly at the undergraduate level, although the book may be useful for introductory graduate courses in non-technical fields as well. Students in programs such as data science, informatics, business, computer science, engineering, information science, biology, physics, statistics, and social sciences will benefit from courses based on this textbook. Their interest will be piqued enough to study network science in greater depth, and perhaps they will choose a career that might land them a job at Google, Facebook, Twitter, or a network start-up of their own.

Pedagogy

No technical mathematical or programming background is required, making the book feasible for introductory courses at any level, including network literacy and programming literacy courses. Such courses may skip the math boxes. By working through the programming tutorials in a collaborative computing lab and assigning the coding exercises, instructors will enable students to acquire sufficient technical skills to perform data-analytic tasks involving networks. This is our approach at Indiana University, where we teach the material in the book over two courses: a first introductory class aimed at sophomore/junior students who have taken or are taking concurrent programming courses in Python; and a second class aimed at junior/senior students. The first course roughly covers the material in Chapters 0–4. The second focuses on Chapters 5–7 after an extended review and some more advanced tutorials on the earlier material.

The extensive programming tutorials and exercises will allow instructors to easily lead and assign hands-on activities, and empower students to reinforce and test their understanding of network concepts. The activities include tutorials on *NetworkX*, a widely adopted library for network analytics; and on all topics covered in the book, from basic exercises to advanced techniques. For example, one tutorial guides students through the steps of extracting social network data from the Web. Using the Twitter application programming interface (API), students will be able to analyze popular topics, identify influential users, and reconstruct information diffusion networks showing how hashtags spread online. Students who carry out the programming tutorials and exercises will become proficient in building, importing/exporting, analyzing, manipulating, and visualizing networks of any type.

The tutorials are in Python, which is the most popular scripting/programming language. A primer that reviews the basic concepts of programming in Python is included in Appendix A. All tutorials are available online as Ipython Notebooks. Over time, NetworkX (and even Python) may evolve and some of the code in the book may need to be updated. We will note such updates in the book's GitHub repository.

There are, of course, other libraries for programming networks, for example *igraph*, *SNAP*, and *graph-tool*. Our selection of NetworkX is based on the fact that it is written in pure Python, which makes it easy to debug for students who are familiar with Python. Many alternatives have Python interfaces but are written in C, making them more efficient but also harder to debug.

Finally, some chapters leverage interactive models to demonstrate network phenomena such as giant components, small worlds, PageRank, preferential attachment, and epidemics. These models run in NetLogo, a popular simulation platform. A tutorial on NetLogo and some of the most relevant models is presented in Appendix B.

About the Cover

The network on the cover, generated by Onur Varol (Ferrara *et al.*, 2016), depicts the diffusion of the #SB277 hashtag on Twitter. This hashtag refers to a 2015 California law on vaccination requirements and exemptions, and the network represents the discussion that was taking place online among supporters and opponents of the bill. Nodes represent Twitter users, and links show information spreading among users via retweets. Node size represents account influence (how many times a user is retweeted) and node colors represent bot scores: red nodes are likely to be bot accounts, blue nodes are likely to be humans.

Acknowledgments

The initial idea for this book originated from conversations with our former colleague Alex Vespignani. Over the years, our colleagues Sandro Flammini, YY Ahn, and Filippo Radicchi have provided precious advice. Several students have assisted in the teaching of our network science courses at Indiana University. Among them, we want to acknowledge Mike Conover, who first conceived some of the exercises found in these pages. We are also grateful to colleagues who provided feedback on early drafts of the book, especially Claudio Castellano, Chato Castillo, and several anonymous reviewers.

We thank our wonderful collaborators, students, postdocs, and visitors: Ana Maguitman, Ben Markines, Bruno Gonçalves, Chengcheng Shao, Diep Thi Hoang, Dimitar Nikolov, Emilio Ferrara, Giovanni Luca Ciampaglia, Jacob Ratkiewicz, Jasleen Kaur, Jose Ramasco, Kai-Cheng Yang, Karissa McKelvey, Kazu Sasahara, Le-Shin Wu, Lilian Weng, Luca Maria Alello, Mark Meiss, Markus Jakobsson, Mihai Avram, Nicola Perra, Onur Varol, Pik-Mai Hui, Prashant Shiralkar, Przemek Grabowicz, Ruj Akavipat, Xiaodan Lou, Xiaoling Sun, Zoher Kachwala, and too many others to name. They are an amazingly bright and fun bunch of people, who have contributed in many ways to the ideas, datasets, and illustrations in the book.

Our work would not be possible without the support of many dedicated staff members in the Center for Complex Networks and Systems Research, the School of Informatics, Computing, and Engineering, and the Indiana University Network Science Institute. Above all, we gratefully acknowledge Tara Holbrook, Michele Dompke, Rob Henderson, Dave Cooley, Patty Mabry, Ann McCranie, Val Pentchev, Matthew Hutchinson, Chathuri Peli Kankanamalage, and Ben Serrette. Thanks also to Nick Gibbons at Cambridge University Press for his encouragement and feedback.

We are grateful to Aric Hagberg, Pieter Swart, and Dan Schult, the authors of NetworkX; and to Uri Wilenski and the Center for Connected Learning and Computer-Based Modeling at Northwestern University for developing and maintaining NetLogo.

Finally, we owe a huge debt of gratitude to our families who love, support, and bear with us even as we work more than we should.

0 Introduction

net·work: (*n.*) an interconnected or interrelated chain, group, or system.

Imagine a world where people have no friends. Where roads never intersect. Where computers are not interconnected. This world without networks would be a very sad and boring place, where nothing happens — and even if something happened, nobody would know. Such a world is unimaginable, because our life is completely defined by networks: relationships, interactions, communications, and the Web. Biological networks governing the interactions between genes in our cells determine our development, neural networks in our brain make us think, information networks guide our knowledge and culture, transportation networks allow us to move, and social networks sustain our life.

Networks are a general yet powerful way to represent and study simple and complex interactions. This book explores the study of networks and how they help us understand the patterns of connections and relationships that shape our lives. In essence, a network is the simplest description of a set of interconnected entities, which we call *nodes*, and their connections, which we call *links*. The network representation is so general and powerful because it strips out many details of a particular system and focuses on the interactions among its elements. Networks are thus used to study widely diverse systems. Nodes can represent all sorts of entities: people, cities, computers, websites, concepts, cells, genes, species, and so on. Links represent relationships or interactions between these entities: friendships among people, flights between airports, packets exchanged among computers on the Internet, links between Web pages, synapses between neurons, and so on.

Before we introduce the basic concepts, definitions, and nomenclature about networks, let us explore a few examples of social, infrastructure, information, and biological networks. Data for all the examples presented here is available on the book's GitHub repository.[1] The networks on which we focus in this book tend to be large, even though one can learn a lot from studying smaller systems, such as social networks built from surveys or interviews. In these cases it is meaningful to examine individual nodes and connections in great detail, whereas analyses of large networks tend to focus on macroscopic properties, classes of nodes and links, typical behaviors, and anomalies.

---

[1] github.com/CambridgeUniversityPress/FirstCourseNetworkScience

0.1 Social Networks

A social network is a group of people connected by some type of relationship. Friendship, collaboration, romance, or mere acquaintance are all examples of social relationships that connect pairs of people. When we talk about a social network, we typically think of a particular type of relationship. A person is represented by a node in the social network, and the relationship is represented by a link between two people. The network is therefore a representation of the relationship. It allows us to talk about the relationship, to describe it and analyze it at a level that goes beyond a pair of people.

There are many different types of social networks, and they are important to study. Health workers analyze networks of sexual relationships to find ways to combat the spread of sexually transmitted diseases. Economists study job referral networks to address inequality and segregation in labor markets. And scientists inspect coauthorship networks in scholarly publications to identify influential thinkers and ideas.

These days we use online social networking sites to keep track of our social ties. Platforms like Facebook and Twitter allow us to keep in touch with many people — partners, friends, colleagues, and acquaintances, sometimes in the hundreds — and communicate with them conveniently, irrespective of distance. Figure 0.1 illustrates a familiar network, a portion of the Facebook social graph. In this network, nodes are people with a Facebook account at a US university, and connections may represent different types of relationship, from real friendship to mere acquaintance. Just looking at the network visualization reveals something about the underlying social structure. Some people have more connections; we represent this by making the corresponding nodes larger and darker. These might be popular students, teachers, or administrators. We also notice that the network is roughly divided into two parts. The data is anonymized so we cannot tell for sure, but a possible interpretation is that the larger subnetwork comprises mostly undergraduate students, and the smaller one includes mostly graduate students. There are connections between nodes in the two groups, but not as many as among nodes within each group. In other words, undergraduate students are more likely to be friends with other undergraduates than with graduate students. Later we will introduce formal names for all these observations, which are typical of most social networks.

The availability of data from online social networks is very exciting for scientists. We can study human interactions at a scale and resolution that was never possible in the past: who befriends whom, who pays attention to what, who likes what, what gets recommended, and how this information propagates through the network. This data provides us with an unprecedented opportunity to discover, track, mine, and model what people do. Like the telescope gave us a first view of distant planets and stars, and the microscope allowed us to peek into living tissue and micro-organisms, social media are enabling the study of social systems and human activity. However, as exciting as these opportunities are to researchers, they don't come without risks of abuse. Online interactions expose our private personal information. We've all heard stories about employers finding embarrassing pictures of prospective employees, or scandals related to hackers and political organizations

Fig. 0.1 Visualization of a network of Facebook users at Northwestern University. Nodes represent people, and links stand for Facebook friend connections.

amassing data about millions of users. The dangers can be subtle. Knowing a little bit of information about a large number of people can reveal a lot more than intended. Using data from Facebook, two MIT students found that just by looking at the gender and sexuality of a person's online friends, they could predict whether that person was gay. Online social networks also make impersonation easy to do and hard to detect. Social phishing is the technique of impersonating a victim's friend (as inferred from an online social network) to induce the victim to disclose sensitive information. Two Indiana University students demonstrated that they were able to obtain the secret passwords of 72% of victims in this way.

Data about a social network can be extracted from many sources. If we want to map human mobility patterns to improve urban transportation networks, we can collect call

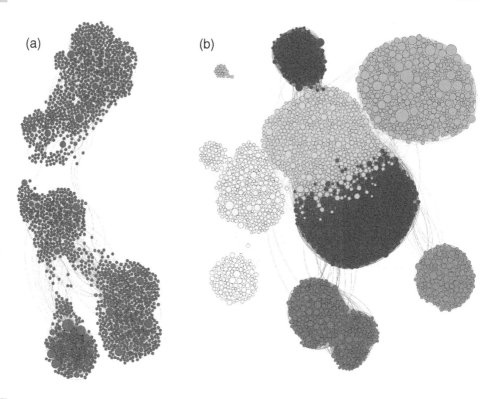

Fig. 0.2 (a) A movie-star network, based on a small sample of movies, actors, and actresses from the Internet Movie Database. Nodes represent movies (blue) or actors/actresses (red). A link connects an actor or actress to a movie in which they starred. (b) A movie co-star network, based on a small sample of actors and actresses from the Internet Movie Database. A link connects two people who have co-starred in at least one movie. Colors represent film genres or languages/countries.

data from cell phones. If we want to map coauthorship among scientists, we can extract the names from a database of scientific publications; two coauthors of the same paper will be linked to each other. (This is not a trivial exercise, because several scientists may have common names.) If we want to map the collaboration among movie stars, we can extract movie credits data from the Internet Movie Database (IMDB.com). Figure 0.2 illustrates two such networks. In one case, there are actually two kinds of nodes: movies and actors/actresses. We draw a link between an actress and a movie in which she has starred. In the other case, we focus on links between actors/actresses who have co-starred in movies. Although the depicted networks capture only tiny portions of the movie database, we again notice some clear patterns. Larger nodes have more connections, representing stars who acted in many movies. We also see that the networks are structured into several dense groups associated with periods, languages, or film genres: Hollywood (blue), Western (cyan), Mexican (purple), Chinese (yellow), Filipino (orange), Turkish and Eastern European (green), Indian (red), Greek (white), and adult (pink) stars in Figure 0.2(b). In Chapter 6 you will learn how to discover these groups and find out what they are about.

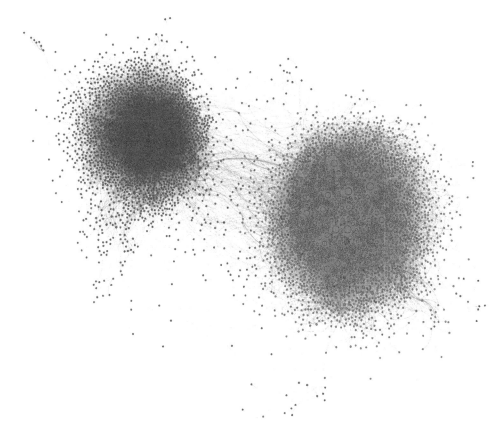

Fig. 0.3 A retweet network on Twitter, among people sharing posts about US politics. Links represent retweets of posts that used hashtags such as `#tcot` and `#p2`, associated with conservative (red) and progressive (blue) messages, respectively, around the 2010 US midterm election. When Bob retweets Alice, we draw a directed link from Alice to Bob to indicate that a message has propagated from her to him. The direction of the links is not shown.

0.2 Communication Networks

In the Facebook and movie networks, links are reciprocal: you cannot friend someone on Facebook unless they agree, and you cannot star in a movie without being listed in the credits. Not all social networks have reciprocal links, however. For example, Twitter is a popular social network with links that are not necessarily reciprocal: Alice can follow Bob without Bob necessarily following Alice back. As a result, the relationships captured by the Twitter network are not friendship; you follow someone to see what they post. When you retweet a post, your followers see it. This is a good way to share information broadly, so Twitter is a social network mainly aimed at spreading information — a communication network. The retweet network in Figure 0.3 illustrates the spread of political messages during a US election. Larger nodes are those with more outgoing links, because how many times users are retweeted by others is a way to measure their influence. You probably

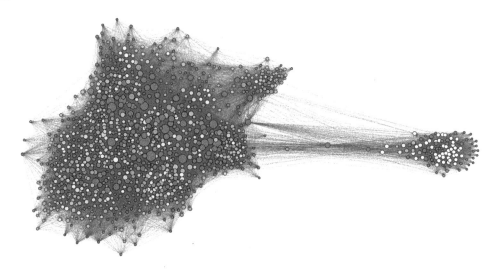

Fig. 0.4   A network based on a database of emails generated by employees of the Enron energy company. The data was acquired by the US Federal Energy Regulatory Commission during its investigation after the company's collapse in 2001. At the conclusion of the investigation, the emails were deemed to be in the public domain and made publicly available for historical research and academic purposes. Only a small portion of the central core of the network is shown. The direction of the links is shown by arrows.

noticed immediately a more striking pattern: conservative users (red nodes) mostly retweet messages from other conservatives, while progressive users (blue nodes) similarly share progressive content. In fact, such preferential patterns of social connections allow us to guess a person's political leaning with high accuracy. This property, called *homophily*, will be discussed in Chapter 2; the algorithm for inferring political preference from the network's structure will be presented in Chapter 6.

Networks like Twitter let us trace the diffusion of hashtags and news, observing how ideas and cultural concepts spread from person to person. But social media are also used to spread misinformation, which is unknowingly passed on by gullible users. Using fake news sites and automated or semi-automated accounts known as "social bots," a malicious entity can cheaply and effectively generate and amplify a disinformation campaign, either for political purposes or to monetize traffic through ads. In recent years we have observed a sharp increase in these types of manipulation of social media on a global scale. If one can control what information people see online, one can manipulate their opinions. This is a threat to democracy in many countries, because without well-informed voters one cannot have free elections. Academic researchers and industry engineers are working hard to develop countermeasures. Understanding the structure and dynamics of the networks that enable the spread of information is a critical component of these efforts.

The social links in Twitter are in place before a user generates a post, which is typically broadcast to all of the user's followers. In email, just like in social networks, nodes are people. However, each message is intended for one or more specific recipients. Links are based on the messages exchanged. Email does not depend on a particular platform; the protocol is open and distributed, so that no single organization controls all of the traffic. As a result, email is still among the most widely used communication networks. Figure 0.4 illustrates an example of an email network. Again, links are directed from the sender to the receiver of an email, indicated by arrows. Node size and color represent two different features: number of incoming and outgoing links, respectively. A larger node receives emails from more people, and a darker node sends emails to more people. The fact that larger nodes tend to be darker and vice versa tells us that there is a correlation between sending and receiving emails.

0.3 The Web and Wikipedia

The Web is the largest information network. While it is now used to provide all kinds of services, it was originally just a network of documents (pages) connected by "hyperlinks," or clickable links. In the early 1990s, Tim Berners-Lee wanted to simplify access by scientists to information about high-energy physics experiments at the European Organization for Nuclear Research (CERN) near Geneva. He came up with three key ideas: (1) a naming system for pages, the Uniform Resource Locator (URL); (2) a simple language for writing documents, called HyperText Markup Language (HTML), including hyperlinks from one page to another; and (3) a simple protocol called HyperText Transfer Protocol (HTTP) for clients (browsers) to talk to servers. With these three components, the Web was born. Berners-Lee even implemented the first Web server and browser software to download pages and media from servers by clicking on links. We can actually see two networks at play here: the static "link graph" made up of a snapshot of Web pages and links at a given time, and the dynamic traffic network emerging from people navigating the Web. To paraphrase the classic philosophical riddle, if there is a link between two pages but nobody clicks on it, is it really part of the Web? The answer of course depends on which of the two networks we are thinking about when we say "Web." In later chapters we will spend more time exploring both of these information networks.

The Web is too large to visualize even a small portion of it in a meaningful way. Let us focus on Wikipedia, which is a network of pages (articles) on a single website. Wikipedia is a collaborative encyclopedia edited by thousands of volunteers around the world, and it is one of the most popular destinations on the Web. There are versions of Wikipedia in many languages, so let us focus on the English one. Still, the English Wikipedia is a huge network with millions of articles (and growing!). So let us focus on just a small subset of articles about math, shown in Figure 0.5. Here, node size represents *PageRank*, a measure of centrality that captures how important an article is based on other articles that link to it — something we will discuss in Chapter 4. For example, the large white node in the middle is the general article about *Mathematics*. Another feature of this network is the presence of a large "core" (gray) and several smaller groups. These groups are tightly connected clusters of articles on specific topics or branches of math. For example, articles

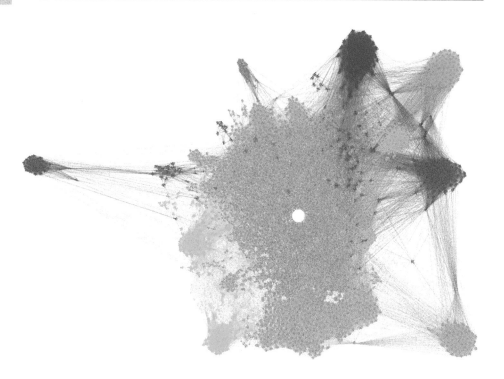

A portion of the Wikipedia information network. Nodes are articles about math. We only consider links among Wikipedia articles, and disregard links to external pages. Node size is proportional to article importance, and colors highlight communities discussed in the text.

about historical Greek (blue), Arab (green), and Indian (brown) mathematicians; about contemporary Indian mathematicians (tan); about math and art (orange), statistics (cyan), game theory (yellow), mathematical software (purple), and pedagogical theory (red). We also observe several "bridge" nodes that connect multiple clusters. These features are found in many real-world networks.

0.4 The Internet

We often think of the Internet as a network of computers and other connected devices, but in reality it is a *network of networks*. In fact the word originates from *internet-working*, or connecting different computer networks through special nodes called *routers*. We can therefore observe the Internet at many levels: at the lowest level we have hard-ware devices that connect individual computers in the same local or wide-area network. These networks are connected by routers, so we can zoom out and think of the network of routers. If we zoom out further we find groups of networks managed by an Inter-net Service Provider (ISP). This organization decides its internal network topology (how

Fig. 0.6 A portion of the Internet router network. The map is a snapshot generated by the Center for Applied Internet Data Analysis (CAIDA.org) using tools that send out small packets of data (probes) between Internet hosts. Colors are assigned according to a community detection algorithm that identifies dense clusters reflecting the geographic distribution of routers. In Chapter 6 you will learn how to use this methodology to study what those clusters represent.

routers are connected) autonomously, and therefore is also called an "autonomous system" (AS). Special "border" routers connect one AS to another, forming what we call the AS network.

Figure 0.6 shows a small portion of the Internet router network. Although the Internet has evolved without central control or coordination, ISPs follow local rules on how to connect their routers. They try to provide the best service at the lowest cost. Certain regularities emerge as a result. For instance, the portion of the Internet that carries the most traffic is often referred to as the "backbone." The large telecommunication companies that manage the Internet backbone have a significant interest in preventing disruption, so they engineer their networks with a lot of redundancy. We thus observe a dense "core," with large routers connected to each other. As we move toward the "periphery" of the Internet — our home routers — the network is more sparsely connected. Such a hierarchical *core–periphery structure* is common in many different types of networks, and will be discussed in Chapter 2. In the router network depicted in Figure 0.6, the green cluster on the left appears well separated from the rest of the network. This is likely due to a bias in the probe methodology used to map these networks: most measurements were taken from the United States, and the routers in this cluster are located there. A related peculiarity is the presence of very large nodes in the green cluster, indicating routers with many connections. This may actually be a measurement error resulting from the same bias. In fact, a router can only have a limited number of connections due to hardware constraints. Let it serve as a reminder that if we use a flawed method to collect data about a network, its analysis may lead to wrong conclusions.

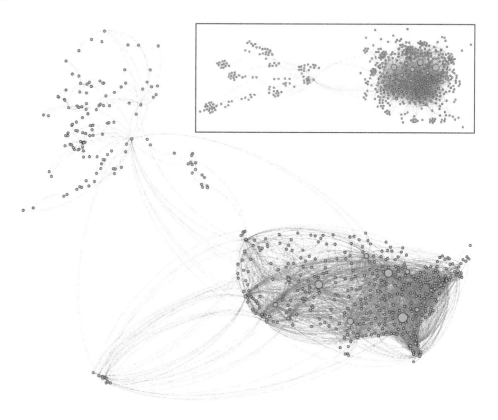

Fig. 0.7   The US air transportation network (flight data from OpenFlights.org). Nodes are positioned according to the geographic coordinates of the corresponding airports, so that we can make out the shape of the continental United States, Alaska, and Hawaii. Note that the map projection makes Alaska appear bigger than its actual size due to its latitude. The airport hubs with most connections (e.g. Atlanta, Chicago, Denver) are clearly recognizable. The inset maps the same network, but with a different "force-directed" layout, discussed in Section 1.10.

0.5 Transportation Networks

Another important class of networks concerns various types of transportation. Nodes are locations: cities, road intersections, airports, ports, train or subway stations. These networks are very different from one another, however. Road networks, for example, evolve in a local fashion to minimize the distance traveled between nearby cities. This leads to the emergence of grid-like structures, in which most nodes have a comparable number of connections — say, four-way intersections. Figure 0.7 shows an air transportation network, which does not have a grid structure. The reason is that airlines try to minimize the number of hops between source and destination without adding costly direct flights between low-traffic airports. The simple solution is to add flights connecting airports to existing hubs. As a result, air flight networks display a "hub and spoke" structure: a few hubs have huge numbers of links, while the majority of nodes have very few connections.

When studying certain types of networks, especially related to transportation and communications, we can think of them in terms of their static structure, or the dynamic processes that occur on these networks. Consider the air transportation network, for instance. We might view the picture in Figure 0.7 as a set of routes that exist between airports, independently of the actual travel that takes place on them; or as a traffic network that emerges from people moving between airports. In the latter sense, links are diverse because they carry different amounts of traffic, and they also change over time. Both the structure and dynamics of networks are important. Sometimes we simply capture the dynamics by representing traffic through link directions and weights, as we discuss in Chapter 4. Other times we may wish to study the actual processes that allow a network to grow and change over time, or the interactions that take place on a network. Chapters 5 and 7 are dedicated to these topics about network dynamics.

0.6 Biological Networks

Within the cells inside our bodies, special molecules called proteins interact in a variety of ways. For example, when a protein folds, its change in structure can regulate the function of another protein or the activity of an enzyme. Enzymes (themselves proteins) catalyze biochemical reactions and are vital to metabolism, which maintains life by harvesting energy for building and supporting the proteins that make up our tissues and organs. Proteins also regulate cell signaling and immune responses. All of these interactions can be seen as networks: protein interaction networks, metabolic networks, gene regulatory networks,

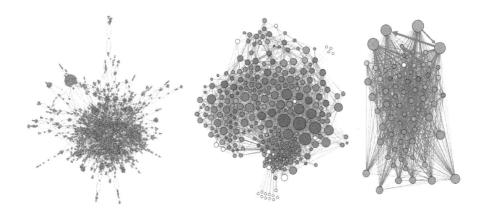

Fig. 0.8   Three biological networks. Left: Protein interaction network of yeast. Node size is proportional to the number of interacting proteins. Center: Neural network of the roundworm *Caenorhabditis elegans*. Large and red nodes represent neurons with more outgoing and incoming synapses, respectively. Right: Food web of species in the Florida Everglades. A directed link goes from a prey to a predator species. The weight (width) of a link represents the energy flux between the two species. Node size and color represent incoming and outgoing links, respectively, so that large blue nodes are the species at the top of the food chain, while small red nodes are the species at the bottom.

and so on. These biological networks exist within a cell. At a higher level, within a body, connections between neural cells (synapses) give rise to the neural networks that form our brains. And at an even higher level, entire species interact. An animal of one species may see another species as food, creating an ecological network, or food web among species. When we think of this network, ecological balance depends on the availability of species that sustain each other. Removing a node in such a food web — when a species goes extinct, for example — affects the survival of other parts of the ecosystem network. Figure 0.8 illustrates three types of biological networks: a protein interaction network, a neural network, and a food web. They are all essential elements of life on our planet.

0.7 Summary

Networks are a general way to model and study complex systems with many interacting elements. We have seen several examples of networks. Nodes can represent many different types of objects, from people to Web pages, from proteins to species, from Internet routers to airports. Nodes can have features associated with them beside labels: geographic location, wealth, activity, number of connections, and so on. Links can also represent many different kinds of relationships, from physical to virtual, from chemical to social, from communicative to informative. They can have a direction (like Web hyperlinks and email) or be reciprocal (like marriage). They can all be the same or have different features such as similarity, distance, traffic, volume, weight, and so on.

0.8 Further Reading

The use of networks to graphically represent social relationships among individuals was introduced by Moreno and Jennings (1934), who called these social networks *sociograms*.

Much more recently, studies have shown that online social networks can reveal a person's sexual orientation (Jernigan and Mistree, 2009) and facilitate highly effective phishing attacks (Jagatic *et al.*, 2007). Conover *et al.* (2011b) showed that political information diffusion networks on Twitter are very polarized and segregated. As a result we can predict the political leaning of most users with high accuracy by starting with a few node labels and propagating them through network neighbors (Conover *et al.*, 2011a).

You can read about the vision, design, and history of the Web in a book coauthored by its inventor (Berners-Lee and Fischetti, 2000).

Spring *et al.* (2002) explain how probes are used to measure the topology of the Internet. Achlioptas *et al.* (2009) show that these approaches have sampling bias. Computer scientists analyze the structure of routers and autonomous system networks to develop models called "topology generators," which can help in the design of these networks (Rossi *et al.*, 2013). To learn more about Internet networks, we recommend the book by Pastor-Satorras and Vespignani (2007).

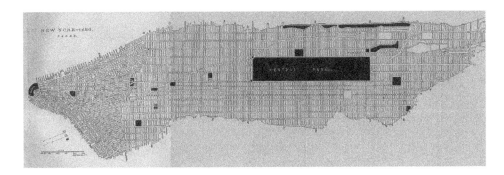

Fig. 0.9   Map of New York in 1880. From Report on the Social Statistics of Cities, Compiled by George E. Waring, Jr., U.S. Census
Office, 1886. Image courtesy of University of Texas Libraries.

Data about the yeast protein interaction network is from Jeong *et al.* (2001). *C. elegans*
neural network data is from White *et al.* (1986). To learn about the human brain network,
or "connectome," we recommend Sporns (2012). The Everglades ecological network is
derived from Ulanowicz and DeAngelis (1998). To learn more about food webs, we refer
to Dunne *et al.* (2002) and Melián and Bascompte (2004).

Data for several of the real-world network examples shown in this book is provided
by the Network Repository (Rossi and Ahmed, 2015). The visualizations are done using
Gephi (Bastian *et al.*, 2009). Layout algorithms are discussed in Chapter 1.

Exercises

0.1   Consider the road map in Figure 0.9. If one were creating a network representation of
traffic patterns, which of the following would be the best choice to make up the links
of the network? (*Hint*: Your answer to the next question may inform your answer to
this question, and vice versa.)
   a. Pedestrians traveling along the streets
   b. Road segments (e.g. 5th Ave. between 12th and 13th streets)
   c. Entire roads (e.g. 5th Ave.)
   d. Vehicles traveling on the roads

0.2   Consider the road map in Figure 0.9. In a network representation of traffic patterns,
which of the following would be the best choice to make up the nodes of the network?
(*Hint*: Your answer to the previous question may inform your answer to this question,
and vice versa.)
   a. City blocks (e.g. the block between 5th–6th avenues and 12th–13th streets)
   b. Street intersections (e.g. 5th Ave. and 12th St.)
   c. Pedestrians moving along the streets
   d. Vehicles traveling on the roads

0.3   Consider the US air transportation network shown in Figure 0.7. Nodes in this network represent airports. What could a link between two airports represent?

0.4   Compare the US air transportation network in Figure 0.7 with the Manhattan road map in Figure 0.9. The air transportation network displays a distinguishing feature that the Manhattan road network lacks. What is this key characteristic?
   a. Singleton nodes with no links
   b. Multiple routes between nodes
   c. Nodes with more than one connected link
   d. Hub nodes with many links

0.5   In a social graph from Facebook, which type of link best represents the "friend" relation? Directed or undirected?

0.6   In a social graph from Twitter, which type of link best represents the "follower" relation? Directed or undirected?

Network Elements

node: (*n.*) a point in a network or diagram at which lines or pathways intersect or branch.

Having seen several examples of real networks in Chapter 0, let us now learn about the basic definitions and quantities that allow us to describe a network.

1.1 Basic Definitions

In very general terms a network, or graph, is a set of elements, which we call *nodes*, along with a set of connections between pairs of nodes, which we call *links*. The links represent the presence of a relation among the elements represented by the nodes. As we have seen earlier, links can correspond to social, physical, communication, geographic, conceptual, chemical, biological, or other interactions. We say that two nodes are *adjacent* or *connected* if there is a link between them. It is also common to call connected nodes *neighbors*.

Networks provide a general theoretical framework allowing for a convenient conceptual representation of interrelations in a wide array of systems; we have seen several examples in Chapter 0. The study of networks has a long tradition in mathematics, computer science, sociology, and communications research. Recently, networks have also been studied intensely in physics and biology. Different fields concerned with networks often introduce their own nomenclature. For example, in some fields a network is called a *graph*,

---

Box 1.1 — **Definition of a Network**

A network G has two parts, a set of N elements, called *nodes* or *vertices*, and a set of L pairs of nodes, called *links* or *edges*. The link (i, j) joins the nodes i and j. A network can be *undirected* or *directed*. A directed network is also called a *digraph*. In directed networks, links are called *directed links* and the order of the nodes in a link reflects the direction: the link (i, j) goes from the source node i to the target node j. In undirected networks, all links are bi-directional and the order of the two nodes in a link does not matter. A network can be *unweighted* or *weighted*. In a weighted network, links have associated *weights*: the *weighted link* (i, j, w) between nodes i and j has weight w. A network can be both directed and weighted, in which case it has directed weighted links.

a node is referred to as a *vertex*, and a link is an *edge*. (We will occasionally use these terms.) The rigorous language for the description of networks is found in graph theory, a field of mathematics that can be traced back to the pioneering work of Leonhard Euler in the eighteenth century. Here we do not want to provide a rigorous introduction to graph theory. We are mostly interested in building a vocabulary and introducing a set of basic notions that will allow us to take our first steps into the world of networks. However, sometimes a formal notation is helpful. In these cases we will include the formal notation in a shaded area or in a box. For example, a more rigorous definition of a network is provided in Box 1.1. In the following chapters we will introduce additional concepts and definitions as needed to analyze real-world systems.

Each network is characterized by the total number of nodes N and the total number of links L. We call N the *size* of the network because it identifies the number of distinct elements composing the system. The numbers of nodes and links do not suffice in defining a network; we have to specify the way in which the nodes are connected by the links.

There are different types of links, which define different classes of networks. In some networks, such as Facebook (Figure 0.1), the links do not have a direction and we represent them as line segments. We call these networks *undirected*. In other cases, such as Wikipedia (Figure 0.5), links are directed and we represent them as arrows. Networks with directed links are called *directed networks*. We say more about directed networks in Section 1.6 and Chapter 4.

In some cases, such as air transportation networks (Figure 0.7), links have associated weights. These are called *weighted networks*. A network can be both directed and weighted. The email network is an example of a weighted directed network, in which link weights and directions represent communication traffic (number of messages) between nodes. We will

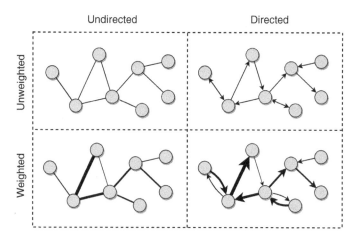

Fig. 1.1 Graphical representations of undirected, directed, and weighted networks. The circles represent the nodes. Pairs of adjacent nodes are connected by a line (link) or arrow (directed link). Arrows indicate the direction of the links. The thickness of a link represents its weight in weighted networks.

return to weighted networks in Section 1.7 and Chapter 4. Figure 1.1 provides illustrations of undirected, directed, and weighted networks.

There are several other classes of networks. A network might have more than one type of node. For example, the movie-star network [Figure 0.2(a)] has two types of nodes representing movies and people. In this network, a link connects an actor or actress to a movie, but there are no links among people or among movies. This is an instance of a so-called *bipartite network*. In a bipartite network, there are two groups of nodes such that links only connect nodes from different groups and not nodes from the same group. Other examples of bipartite networks include those that capture the relationships between songs and artists, between classes and students, and between products and customers. More on bipartite networks in Chapter 4.

A network might have multiple types of links, in which case it is called a *multiplex* network. To use the movie-star example again, we could imagine adding links between actors and/or actresses who are married to each other. In the example of Wikipedia (Figure 0.5), in addition to the hyperlinks, we might have weighted links representing clicks from Wikipedia users, and/or undirected links between articles that share editors. These and other more complex types of networks are discussed further in Section 1.8.

1.2 Handling Networks in Code

To manage, analyze, and visualize networks with more than a handful of nodes and links, we need to use software tools or write our own code. There are many network analysis and visualization tools, as well as libraries to handle networks in many programming languages. Throughout the book we will occasionally mention a couple of these tools. For instance, the visualizations in Chapter 0 are generated with an application called *Gephi*. However, we believe that to get a hands-on understanding of networks it is necessary to "get our hands dirty" and write some code. We assume that students using this book have some familiarity with Python, a popular programming language among both novice and expert coders.[1] To make life easier, we will use *NetworkX* (networkx.github.io), a Python package for the creation, manipulation, and study of the structure, dynamics, and function of networks. NetworkX provides data structures, algorithms, measures, and generators for networks, as well as rudimentary visualization facilities.[2]

Once we import NetworkX, we can easily create an undirected network ("Graph") and add a few nodes and links. Nodes are referred to by integer IDs and links are called edges:

---

[1] We offer an introductory tutorial on Python in Appendix A; it can also be downloaded from the book's GitHub repository at github.com/CambridgeUniversityPress/FirstCourseNetworkScience.

[2] We offer an introductory tutorial on NetworkX on the book's GitHub repository.

```
import networkx as nx # always import NetworkX first!
G = nx.Graph()
G.add_node(1)
G.add_node(2)
G.add_edge(1,2)
```

We can add several nodes or links at once:

```
G.add_nodes_from([3,4,5,...])
G.add_edges_from([(3,4),(3,5),...])
```

Here is how we get lists of nodes, links, and neighbors of a given node:

```
G.nodes()
G.edges()
G.neighbors(3)
```

And here is how you loop over nodes or links:

```
for n in G.nodes:
    print(n, G.neighbors(n))
for u,v in G.edges:
    print(u, v)
```

Similarly, we can create a directed network ("DiGraph"):

```
D = nx.DiGraph()
D.add_edge(1,2)
D.add_edge(2,1)
D.add_edges_from([(2,3),(3,4),...])
```

Note that the link from node **1** to node **2** is distinct from the link from node **2** to node **1** because this network is directed. Also note that when we add a link, the nodes are added automatically if they don't already exist. This is convenient. There are functions for getting the size and number of links:

```
D.number_of_nodes()
D.number_of_edges()
```

In a directed network, when we ask for the neighbors of a node, we get both the nodes linking to and from it. But there are also functions to get only the edges linking to or from, respectively called predecessors and successors:

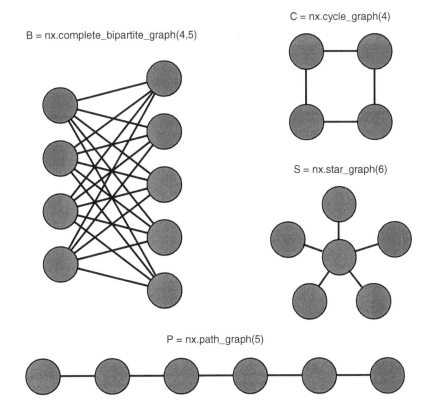

B = nx.complete_bipartite_graph(4,5)

C = nx.cycle_graph(4)

S = nx.star_graph(6)

P = nx.path_graph(5)

Fig. 1.2   A few simple networks generated by NetworkX functions: complete bipartite (B), cycle (C), star (S), and path (P). The concept of a *complete* network is introduced in the next section.

```
D.neighbors(2)
D.predecessors(2)
D.successors(2)
```

Finally, there are functions to generate networks of many types. Typically these functions need arguments that specify the number of nodes or links. Here is code to generate a few networks, shown in Figure 1.2:

```
B = nx.complete_bipartite_graph(4,5)
C = nx.cycle_graph(4)
P = nx.path_graph(5)
S = nx.star_graph(6)
```

We strongly recommend that you read the NetworkX tutorial[3] and bookmark its documentation.[4] And remember, Google and Stack Overflow are your friends when you are stuck!

[3] networkx.github.io/documentation/stable/tutorial.html
[4] networkx.github.io/documentation/stable/

1.3 Density and Sparsity

The maximum number of links in a network is bounded by the possible number of distinct connections among the nodes of the system. The maximum number of links is therefore given by the number of pairs of nodes. A network with the maximum number of links, in which all possible pairs of nodes are connected by links, is called a *complete network*.

The maximum number of links in an undirected network with N nodes is the number of distinct pairs of nodes:

$$L_{max} = \binom{N}{2} = N(N-1)/2. \tag{1.1}$$

Intuitively each node can connect to $N-1$ other nodes, and there are N of them. However, that would count each pair twice, so we divide by two. In a directed network, each pair of nodes should be counted twice, once for each direction, so $L_{max} = N(N-1)$. Counting the possible pairs of objects among a set of N objects is something that we will encounter again later in the book. Mathematicians have a name for the formula $\binom{N}{2}$: "N choose two."

A bipartite network is *complete* if each node in one group is connected to all nodes in the other group (see example B in Figure 1.2). In this case $L_{max} = N_1 \times N_2$, where N_1 and N_2 are the sizes of the two groups.

The fraction of possible links that actually exist, which is the same as the fraction of pairs of nodes that are actually connected, is called the *density* of the network. A complete network has maximal density: one. However, the actual number of links is typically much smaller than the maximum, as most pairs of nodes are not directly connected to each other. Therefore the density is often much, much smaller than one — by orders of magnitude in most real-world, large networks. This is an important feature that helps in dealing with network structure. We call it *sparsity*. Intuitively the fewer edges are in a network, the sparser it is.

The density of a network with N nodes and L links is

$$d = L/L_{max}. \tag{1.2}$$

In an undirected network this is given by

$$d = L/L_{max} = \frac{2L}{N(N-1)} \tag{1.3}$$

and in a directed network the density is

$$d = L/L_{max} = \frac{L}{N(N-1)}. \tag{1.4}$$

Table 1.1 Basic statistics of network examples. Network types can be (D)irected and/or (W)eighted. When there is no label, the network is undirected and unweighted. For directed networks, we provide the average in-degree (which coincides with the average out-degree)

Network	Type	Nodes (N)	Links (L)	Density (d)	Average degree ($\langle k \rangle$)
Facebook Northwestern Univ.		10,567	488,337	0.009	92.4
IMDB movies and stars		563,443	921,160	0.000006	3.3
IMDB co-stars	W	252,999	1,015,187	0.00003	8.0
Twitter US politics	DW	18,470	48,365	0.0001	2.6
Enron email	DW	87,273	321,918	0.00004	3.7
Wikipedia math	D	15,220	194,103	0.0008	12.8
Internet routers		190,914	607,610	0.00003	6.4
US air transportation		546	2,781	0.02	10.2
World air transportation		3,179	18,617	0.004	11.7
Yeast protein interactions		1,870	2,277	0.001	2.4
C. elegans brain	DW	297	2,345	0.03	7.9
Everglades ecological food web	DW	69	916	0.2	13.3

In a complete network, $d = 1$ by definition, since $L = L_{max}$. In a sparse network, $L \ll L_{max}$ and therefore $d \ll 1$. When a network grows very large, we can observe how the number of links increases as a function of the number of nodes. We say that the network is sparse if the number of links grows proportionally to the number of nodes ($L \sim N$), or even slower. If instead the number of links grows faster, e.g. quadratically with network size ($L \sim N^2$), then we say that the network is dense.

To illustrate the importance of network sparsity, let us consider the example of Facebook. At the time of writing, Facebook has around 2 billion users ($N \approx 2 \times 10^9$). If this was a complete network, there would be $L \approx 10^{18}$ links — that is a number with 18 zeros, and there is no way to store so much data! But fortunately, social networks are very sparse and Facebook is no exception. Each user has on average 1000 friends or less, so that the density is approximately $d \approx 10^{-6}$. That is still a lot of data, but Facebook can manage it.

Table 1.1 presents basic statistics about the size and density of the network examples illustrated in Chapter 0.[5] Although these networks are very different from each other, they are all sparse.

NetworkX makes it easy to measure the density of directed and undirected networks:

---

[5] Datasets for these networks are available in the book's GitHub repository: github.com/CambridgeUniversity Press/FirstCourseNetworkScience

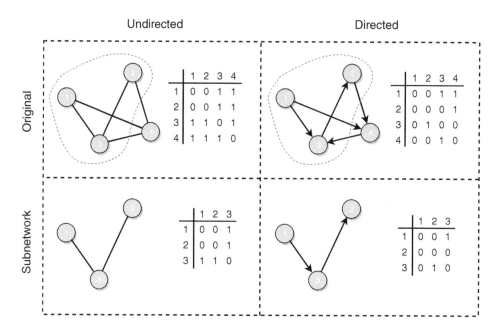

Network and subnetwork examples. We also show the adjacency matrix representation of each network (see Section 1.9).

```
nx.density(G)
nx.density(D)
CG = nx.complete_graph(8471)  # a large complete network
print(nx.density(CG))          # no need for a calculator!
```

1.4 Subnetworks

In many cases, we are interested in a subset of a network, which is itself a network and is called a *subnetwork* (or *subgraph*). A subnetwork is obtained by selecting a subset of the nodes and *all* of the links among these nodes.

Figure 1.3 provides some illustrations of subnetworks of undirected and directed networks. The abundance of certain types of subnetworks and their properties is important in the characterization of real networks. As an example, a *clique* is a complete subnetwork: a subset of nodes all linked to each other. Any subnetwork of a complete network is a clique because all pairs of nodes in the network are connected and therefore all pairs of nodes in any subnetwork are also connected.

A special type of subnetwork is the *ego network* of a node, which is the subnetwork consisting of the chosen node — called the *ego* — and its neighbors. Ego networks are often studied in social network analysis.

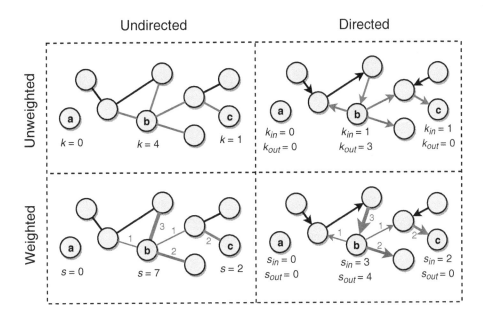

Fig. 1.4 Illustrations of degree and strength in directed, undirected, weighted, and unweighted networks. The links of nodes **a**, **b**, and **c** along with their weights are highlighted in red, and their degrees or strengths are shown.

Using NetworkX we can generate a subnetwork of a given network by specifying a subset of nodes:

```
K5 = nx.complete_graph(5)
clique = nx.subgraph(K5, (0,1,2))
```

1.5 Degree

The *degree* of a node is its number of links, or neighbors. We use k_i to denote the degree of node i. Figure 1.4 illustrates the degree of a few nodes in an undirected network. A node with no neighbors, such as node **a** in the figure, has degree zero ($k = 0$) and is called a *singleton*.

The average degree of a network is denoted by $\langle k \rangle$. It is an important property and is related (directly proportional) to its density.

The average degree of a network is defined as

$$\langle k \rangle = \frac{\sum_i k_i}{N}. \tag{1.5}$$

Since each link contributes to the degree of two nodes in an undirected network, the numerator of Eq. (1.5) can be written as $2L$. From the definition of density for an undirected network [Eq. (1.3)], $2L = dN(N - 1)$. Therefore

$$\langle k \rangle = \frac{2L}{N} = \frac{dN(N-1)}{N} = d(N-1) \qquad (1.6)$$

and conversely

$$d = \frac{\langle k \rangle}{N-1}. \qquad (1.7)$$

This makes sense: the maximum possible degree of a node is $k_{max} = N - 1$, obtained when the node is connected to every other node. Intuitively, the density is the ratio between the average and maximum degree.

Table 1.1 shows the average degree of the network examples illustrated in Chapter 0. NetworkX has a function that returns the degree of a given node. Without arguments, it returns a dictionary with the degree of each node:

```
G.degree(2) # returns the degree of node 2
G.degree()  # returns the degree of all nodes of G
```

In Chapter 3 we will see that the degrees of a network's individual nodes are very important properties to characterize the structure of the network. So far we have defined the degree in undirected networks. Next we extend the definition to directed and weighted networks.

1.6 Directed Networks

In the graphical representation of a network, the directed nature of the links is depicted by means of an arrow, indicating the direction of each link. The main difference between directed and undirected networks is represented in Figure 1.1. In an undirected network, the presence of a link between two nodes connects the adjacent nodes in both directions. In contrast, the presence of a link in a directed network does not necessarily imply the presence of a link in the opposite direction. This fact has important consequences for the connectedness of a directed network, as will be discussed in more detail in Chapter 2.

When we consider the degree of a node in a directed network, we have to think of incoming and outgoing links separately. The number of incoming links, or predecessors, of node i is called the *in-degree* and denoted by k_i^{in}. The number of outgoing links, or successors, of node i is called the *out-degree* and denoted by k_i^{out}. Figure 1.4 illustrates the in- and out-degrees of a few nodes in a directed network.

We already defined the density for a directed network [Eq. (1.4)]. We can define average in-degree and average out-degree similarly to Eq. (1.5).

NetworkX has functions that return the in-degree and out-degree of a given node. If the network is directed, the `degree` function returns the total degree, which is the sum of in-degree and out-degree:

```
D.in_degree(4)
D.out_degree(4)
D.degree(4)
```

1.7 Weighted Networks

In the graphical representation of a network, the weighted nature of the links is depicted by means of lines of different width, indicating the weight of each link. A weight of zero is equivalent to the absence of a link. The main difference between weighted and unweighted networks is represented in Figure 1.1.

A weighted network can be directed or undirected; let us first assume the simpler case of an undirected weighted network. We can measure the degree of a node in a weighted network by disregarding the weights. However, it may be important to consider the weights. We can therefore define the *weighted degree*, or *strength* of a node, as the sum of the weights of its links. Similarly, we can define *in-strength* and *out-strength* in the case of a directed weighted network. Both cases are illustrated in Figure 1.4.

> The weighted degree, or *strength*, of node i in an undirected weighted network is denoted by
>
> $$s_i = \sum_j w_{ij}, \tag{1.8}$$
>
> where w_{ij} is the weight of the link between nodes i and j. We assume $w_{ij} = 0$ if there is no link between i and j. We can analogously generalize in-degree and out-degree to in-strength and out-strength in a directed weighted network:
>
> $$s_i^{in} = \sum_j w_{ji}, \tag{1.9}$$
>
> $$s_i^{out} = \sum_j w_{ij}, \tag{1.10}$$
>
> where w_{ij} is the weight of the directed link from i to j.

In NetworkX, both graphs and digraphs can have "weight" attributes attached to links. When adding multiple weighted links, each is specified as a triple where the third element is the weight:

```
W = nx.Graph()
W.add_edge(1,2,weight=6)
W.add_weighted_edges_from([(2,3,3),(2,4,5)])
```

We can get a list of links with their associated weight data, for example if we need to print the links with large weights:

```
for (i,j,w) in W.edges(data='weight'):
    if w > 3:
        print('(%d, %d, %d)' % (i,j,w)) # skip link (2,3)
```

Finally, we can get the strength of a given node using the `degree` function and specifying the weight attribute:

```
W.degree(2, weight='weight') # strength of node 2
                             # is 6 + 3 + 5 = 14
```

1.8 Multilayer and Temporal Networks

In the US air transportation network of Figure 0.7, the links represent direct flights between airports, regardless of which particular airlines operate those flights. But classifying the flights according to their respective airlines is valuable in a number of situations. We may wish to predict the propagation of scheduling delays through an airline's network, or investigate the consequences of such delays on the movement of passengers. In fact, each commercial airline tries to reschedule passengers on its own flights first because it is expensive to rebook them on another company's flights. Therefore the air transportation network of a specific airline has its own identity, even though it is intertwined with the networks of other airlines. In these cases it is beneficial to represent the system as a *multilayer network* (i.e. a combination of layers), where each layer is the air transportation network of a specific airline: the nodes are the airports, the links flights operated by the same company.

If each layer in a multilayer network is built upon the same set of nodes, the network is called a *multiplex*. The air transportation network is an example of a multiplex. Another example is a social network in which the different layers represent different types of social relationships. For example, one layer could represent friendship ties, another layer family ties, another coworker ties, and so on. The nodes in each layer represent the same individuals.

A *temporal network* is a special case of a multiplex. Links are dynamic, in that the respective node–node interactions occur at different times. Nodes may also have a dynamic character, in that they may appear and disappear at different stages of the network evolution. For instance, networks of user activity on Twitter are temporal because posts, retweets,

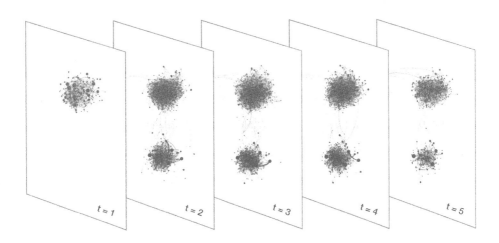

Fig. 1.5   Temporal network of political retweets. Each snapshot includes retweet links with timestamps in a particular time interval. By aggregating these snapshots over time, we obtain the static network shown in Figure 0.3.

and mentions occur at different times, which can be identified by their timestamps. We can divide the time span of a temporal network into consecutive intervals: all nodes and links existing during each interval constitute a *snapshot* of the system. Each snapshot can be interpreted as one layer of a multiplex, as illustrated in Figure 1.5.

In a multilayer network there are *intralayer links*, connecting pairs of nodes in the same layer, and *interlayer links*, connecting pairs of nodes in different layers. In the special case of multiplex networks, interlayer links connect each node of a layer with its counterpart in the other layers. Such links are called *couplings*, because they couple copies of the same node in different layers.

Traditionally, multiplex networks have been analyzed by aggregating data from the different layers and then studying the resulting network. For instance, the networks of Figures 0.3 and 0.7 are aggregations of multiplex networks corresponding to time intervals and different airlines, respectively. The aggregated network is typically weighted, even when the links of the multiplex are not, because there are usually multiple links joining the same pair of nodes in different layers, which turn into a single weighted link in the aggregated system. For example, the links in Figure 0.3 are weighted by the number of times a user retweets another. But aggregation discards a lot of valuable information provided by the original multiplex system. In the air transportation case, merging networks corresponding to different airlines prevents us from studying transitions of passengers between such networks, which may become necessary in case of strikes or technical problems affecting a specific airline.

In general, each layer could be characterized by its own set of nodes and links. Therefore, layers may represent entirely different graphs and the resulting system is a *network of networks*. Here, interlayer links may represent dependency relationships between the nodes of the networks. Consider the electrical power grid, which connects power-generating stations and demand centers through high-voltage transmission lines. The stations are controlled by computers that monitor and manage the production and transmission of electricity. These computers are connected through the Internet. In turn, Internet routers depend on power stations for their electricity supply. Therefore we have a system with two coupled networks: the power grid and the Internet.

In such a coupled system, one network can affect the other to optimize delivery; the grid can be reconfigured to reroute power when needed. However, this kind of network of networks can also introduce unpredictable vulnerabilities. A software problem or attack can take down one or more nodes in the power grid, and without electricity the Internet in an area could also go down, leading to failures of other nodes and, in an extreme case, a catastrophic domino effect called a *cascading failure* affecting a large portion of the grid. For these reasons, networks of networks are the subject of intense study.

To keep things simple, in this book we focus mainly on networks with a single type of node and a single type of link. In an undirected network, we will assume that there can be at most one link connecting a pair of nodes. (If the network is directed, there can be two links, one in each direction, as shown in Figure 1.1.) In addition, we will not consider *self-loops*, or links connecting a node to itself; we will assume that each link connects two distinct nodes.

1.9 Network Representations

To store/retrieve a network in/from a computer file or memory, we need a way to formally represent its nodes and links. There are several possible network representations. The simplest is the *adjacency matrix*, an $N \times N$ matrix in which each element represents the link between the nodes indexed by the corresponding row and column.

> Element a_{ij} of the adjacency matrix represents the link between nodes i and j. $a_{ij} = 1$ if i and j are adjacent, $a_{ij} = 0$ otherwise.

In Figure 1.3 we show the graphical illustrations of different undirected and directed networks and their corresponding adjacency matrices.

For undirected networks the adjacency matrix is symmetric: we can swap rows and columns and the matrix does not change. Therefore, half of the matrix contains redundant information. For directed networks, the adjacency matrix is not symmetric. For unweighted networks, the elements take only values one or zero to indicate the presence or absence of a link, respectively. For weighted networks, matrix elements can take any values corresponding to the link weights. We have already encountered the adjacency matrix elements for weighted networks [w_{ij} in Eqs (1.8)–(1.10)].

In NetworkX, we can get and print adjacency matrices and use the matrix representation to get and set link attributes:

```
print(nx.adjacency_matrix(G)) # graph
G.edge[3][4]
G.edge[3][4]['color']='blue'
```

```
print(nx.adjacency_matrix(D)) # digraph
D.edge[3][4]
D.edge[4][3] # not the same as the previous one
print(nx.adjacency_matrix(W)) # weighted graph
W.edge[2][3]
W.edge[2][3]['weight'] = 2
```

While the adjacency matrix representation matches the mathematical formalism of networks, it is not efficient for storing real networks, which are typically large and sparse. The required storage space grows like the square of the network size (N^2), but if the network is sparse, most of this space is wasted storing zeros (non-existing links). With large sparse networks, a more compact network representation is the *adjacency list*, a data structure that stores the list of neighbors for each node. Adjacency lists represent sparse networks efficiently because the non-existing links are ignored; only the existing links (non-zero values of the adjacency matrix) are considered.

NetworkX provides facilities to loop over a network's adjacency list and retrieve links and their attributes. For example, here is one way to print the neighbors of each node:

```
for n,neighbors in G.adjacency():
    for number,link_attributes in neighbors.items():
        print('(%d, %d)' % (n,number))
```

A third, equally efficient network representation is the *edge list*, which lists each link as a pair of connected nodes. We may also need to list the nodes separately in case of singletons, which would not appear in any of the pairs. In the case of weighted networks, each link is represented as a triple, where the third element is the weight.

In this book we will use the edge list representation to store networks. NetworkX has functions to write and read network files using this representation. You can view the format of an edge list file for yourself:

```
nx.write_edgelist(G, "file.edges")
G2 = nx.read_edgelist("file.edges")         # G2 same as G
nx.write_weighted_edgelist(W, "wf.edges") # store weights
with open("wf.edges") as f:
    for line in f:
        print(line)
W2 = nx.read_weighted_edgelist("wf.edges") # W2 same as W
```

1.10 Drawing Networks

We can learn a lot about a network by drawing it and inspecting its graphical representation. This requires a *network layout algorithm* to place each node on a plane. (There are also sophisticated 3D layouts, but we do not discuss them in this book.) There are many layout algorithms that are appropriate for representing different kinds of networks; for example, we used a *geographic layout* to draw the air transportation network in Figure 0.7. For relatively small networks, layouts that place nodes along concentric circles or layers can reveal important hierarchical structure. The most popular class of network layout algorithms are *force-directed layout algorithms*, which are used to visualize most of the example networks in Chapter 0. The inset of Figure 0.7 uses a force-directed layout as well.

The goals of a force-directed layout algorithm are to place the nodes so that connected nodes are positioned close to each other, all the links are of similar length, and the number of link crossings is minimized. To get an idea of how force-directed layout works, imagine a force that repels any two nodes from each other, like the force between two particles with the same electrical charge. Further imagine a spring connecting any two linked nodes, generating an attractive force when they are too far from each other. Force-directed layout algorithms simulate such a physical system so that nodes move to minimize the energy of the system: connected nodes will move toward each other and away from nodes not connected to them.

The result is not only an aesthetically pleasing drawing, but also, sometimes, a visualization of the most obvious communities in the network, as we have seen in Chapter 0. For example, in Figure 0.3, because people in a community (progressive or conservative) are densely connected to each other, they end up clustered together in the layout.

NetworkX has a function to draw a network, which uses a rudimentary network layout algorithm:

```
import matplotlib.pyplot
nx.draw(G)
```

Note that drawing requires a plot interface, such as Matplotlib's. This works reasonably well for small networks with, say, less than 100 nodes. For larger networks, there are better visualization tools. The examples in Chapter 0 are visualized with Gephi's *ForceAtlas2* layout algorithm.

1.11 Summary

We have presented some basic definitions and quantities that allow us to describe a network:

1. A network is made up of two sets of elements: the nodes and links connecting pairs of nodes.
2. A subnetwork is a subset of the network including some of its nodes and all of the links among them.
3. In directed networks, links have a direction. There may be a link from node **1** to node **2**, and not necessarily one from node **2** to node **1**. In undirected networks, links are reciprocal.

4. In weighted networks, links have associated weights that represent connection attributes like importance, similarity, distance, traffic, etc. In unweighted networks, all links are the same.
5. Multilayer networks have different types of nodes and links, divided into interconnected layers. If the nodes are the same in each layer, the multilayer network is called a multiplex.
6. The density of a network is the fraction of node pairs that are connected. A network is complete if all pairs of nodes are connected, so that the density is one. Most real networks are sparse, meaning that they have very small density.
7. The degree of a node is the number of neighbors. In directed networks, nodes have in-degree and out-degree measuring the number of incoming and outgoing links, respectively. If the network is weighted, the strength of a node is the sum of the weights of its links. The nodes of weighted directed networks have in-strength and out-strength.
8. Adjacency lists and edge lists are efficient representations to store sparse networks.
9. NetworkX is a popular and convenient programming library to code networks in the Python language.

The definitions in this chapter form a basic vocabulary for network science. More quantities and properties will be introduced in future chapters so that we can describe, analyze, and model real networks and learn what they tell us about the underlying systems and phenomena.

1.12 Further Reading

There are several other excellent textbooks on network science to go beyond the introductory material in this book. Caldarelli and Chessa (2016) dig a bit deeper into the data science of several case studies. If you are interested in branching into physics, consider the textbook by Barabási (2016); if you want to explore the connections to economics and sociology, we recommend the textbook by Easley and Kleinberg (2010). For more advanced physics, math, and social science topics, there are many books to choose from (Wasserman and Faust, 1994; Caldarelli, 2007; Barrat *et al.*, 2008; Cohen and Havlin, 2010; Bollobás, 2012; Dorogovtsev and Mendes, 2013; Latora *et al.*, 2017; Newman, 2018).

Kivelä *et al.* (2014) and Boccaletti *et al.* (2014) have provided influential reviews on multilayer networks. Temporal networks are reviewed by Holme and Saramäki (2012). Gao *et al.* (2012) analyze networks of networks. Catastrophic failure in these networks is discussed by Reis *et al.* (2014) and Radicchi (2015).

For background on network drawing, see Di Battista *et al.* (1998). Force-directed network layout (also known as spring layout) algorithms were introduced by Eades (1984) and improved by Kamada and Kawai (1989) and Fruchterman and Reingold (1991). The ForceAtlas2 layout algorithm, used for many visualizations in this book, was developed by Jacomy *et al.* (2014).

Exercises

1.1   Go through the Chapter 1 Tutorial on the book's GitHub repository.[6]

1.2   Consider a network with N nodes. Given a single link, what is the maximum number of nodes that link can connect? Given a single node, what is the maximum number of links that can connect to that node?

1.3   Consider the road map in Figure 0.9. The grid-like structure of this network means that most nodes have the same degree. What is the most common degree for nodes in this network?

1.4   Consider the road map in Figure 0.9. Manhattan has a lot of one-way streets. This implies that a good network model of traffic flow would probably have directed links. Consider a subgraph of this network with grid-like connectivity and all one-way streets (i.e. each node is a four-way intersection of two one-way streets). What is the most common in-degree of nodes in this subgraph? What is the most common out-degree?

1.5   What network quantity can we use to represent the volume of traffic between each pair of adjacent intersections in the Manhattan road map (Figure 0.9)?

1.6   Consider a directed network of N nodes. Now consider the total in-degree (i.e. the sum of the in-degree over all nodes in the network). Compare this to the analogous total out-degree. Which of the following must hold true for any such network?
   a. Total in-degree must be less than total out-degree
   b. Total in-degree must be greater than total out-degree
   c. Total in-degree must be equal to total out-degree
   d. None of these hold true in all instances

1.7   Consider a Twitter retweet network, where users are nodes and we want to show how many times a given user has retweeted another user. What link type best captures this relation?
   a. Undirected, unweighted
   b. Undirected, weighted
   c. Directed, unweighted
   d. Directed, weighted

1.8   Consider a hashtag co-occurrence graph from Twitter. In this network, hashtags are the nodes, and a link between two hashtags indicates how often those two hashtags appear in tweets together. What link type would best capture this relation?
   a. Undirected, unweighted
   b. Undirected, weighted
   c. Directed, unweighted
   d. Directed, weighted

---

[6] github.com/CambridgeUniversityPress/FirstCourseNetworkScience

1.9   Consider a network created from characters in a story or play. The nodes are people, and a link exists between two nodes if those characters ever engage in dialogue. Which type of edge could represent this relation? Justify your answer.
   a. Undirected, unweighted
   b. Undirected, weighted
   c. Directed, unweighted
   d. Directed, weighted

1.10 Suppose we want to make a more complex version of a dialog network that captures how much each character speaks and to whom. What type of link would best represent this relation?
   a. Undirected, unweighted
   b. Undirected, weighted
   c. Directed, unweighted
   d. Directed, weighted

1.11 Imagine that your social network has a subnetwork where you and 24 of your friends (25 people total) are all friends with each other. What is such a subnetwork called? And how many links are contained in the subnetwork?

1.12 Consider an undirected network with N nodes. What is the maximum number of links this network can have?

1.13 Consider a bipartite network of N nodes, N_1 nodes of type 1 and N_2 nodes of type 2 (so that $N_1 + N_2 = N$). What is the maximum number of links in this network?

1.14 Given a complete network A with N nodes, and a bipartite network B also with N nodes, which of the following holds true for any $N > 2$:
   a. Network A has more links than network B
   b. Network A has the same number of links as network B
   c. Network A has fewer links than network B
   d. None of these hold true for all such $N > 2$

1.15 Recall that in a complete network there exists a link between each pair of nodes. We know that a complete undirected network of N nodes has $N(N-1)/2$ edges. Must any undirected network of N nodes and $N(N-1)/2$ links be complete? Explain why or why not.

1.16 Consider this adjacency matrix:

$$
\begin{array}{c}
 \\
A \\ B \\ C \\ D \\ E \\ F
\end{array}
\begin{array}{cccccc}
A & B & C & D & E & F \\
\left(\begin{array}{cccccc}
0 & 1 & 0 & 0 & 0 & 0 \\
0 & 0 & 2 & 0 & 0 & 0 \\
0 & 0 & 0 & 0 & 0 & 0 \\
0 & 1 & 0 & 0 & 1 & 0 \\
0 & 0 & 0 & 0 & 0 & 1 \\
2 & 1 & 3 & 1 & 1 & 0
\end{array}\right).
\end{array}
\qquad (1.11)
$$

An entry in the ith row and jth column indicates the weight of the link from node i to node j. For instance, the entry in the second row and third column is 2, meaning the weight of the link from node **B** to node **C** is 2. What kind of network does this matrix represent?

a. Undirected, unweighted

b. Undirected, weighted

c. Directed, unweighted

d. Directed, weighted

1.17 Consider the network defined by the adjacency matrix in Eq. (1.11). How many nodes are in this network? How many links? Are there any self-loops?

1.18 Consider the network defined by the adjacency matrix in Eq. (1.11). Are there any nodes with outgoing links to every other node? If so, which nodes? Are there any nodes with in-links from every other node? If so, which nodes?

1.19 Consider the network defined by the adjacency matrix in Eq. (1.11). A *sink* is defined as a node with in-links but no out-links. Which nodes in the network, if any, have this property?

1.20 Consider the network defined by the adjacency matrix in Eq. (1.11). What is the in-strength of node **C**? What is its out-strength?

1.21 Convert the network defined by the adjacency matrix in Eq. (1.11) to an undirected, unweighted graph. (When converting a directed graph to an undirected one, nodes i and j are connected in the undirected graph if there is a directed link from i to j, or from j to i, or both.) You may want to print out the resulting matrix and/or draw a network diagram for reference. How many nodes are in this converted network? How many links?

1.22 Consider the unweighted, undirected version of the network defined by the adjacency matrix in Eq. (1.11), constructed as explained in Exercise 1.21. What is the minimum degree in this network? What is the maximum degree? What is the mean degree? What is the density?

1.23 Imagine two different undirected networks, each with the same number of nodes and links. Must both networks have the same maximum and minimum degree? Explain why or why not. Must they have the same mean degree? Explain why or why not.

1.24 We have seen that Facebook's network is incredibly sparse. Assume it has approximately 1 billion users, each with 1000 friends on average.

- Suppose Facebook releases its annual report and it shows that while the number of users in the network has stayed the same, the average number of friends per user has increased. Would this imply that the network density increased, decreased, or stayed the same?

- Suppose instead that both the number of users and the average number of friends per user doubled. Would this imply the network density increased, decreased, or stayed the same?

1.25 Netflix keeps data on customer preferences using a big bipartite network connecting users to titles they have watched and/or rated. Netflix's movie library contains approximately 100,000 titles if you count streaming and DVD-by-mail. In the fourth quarter of 2013, Netflix reported having about 33 million users. Assume the average user's degree in this network is 1000. Approximately how many links are in this network? Would you consider this network sparse or dense? Explain.

1.26 Netflix keeps data on customer preferences using a big bipartite network connecting users to titles. Suppose that from 2013 to 2014 Netflix's library has remained the same size, while the number of users has increased. Further suppose that the average user's degree in this network has remained constant. Has the density of this network increased, decreased, or stayed the same?

## 2

Small Worlds

path: (*n.*) a sequence of arcs in a network that can be traced continuously without retracing any arc.

Many types of networks display a few fundamental features. In this chapter we introduce three of these characteristics: similarity between neighbors, short paths connecting nodes, and triangles formed by common neighbors. *Social networks* provide us with familiar cases to illustrate these features. In a social network, nodes represent people and links represent some type of social relationship, such as friendship, work, acquaintances, or family ties. Social networks are the most extensively studied category of networks; there is a century of literature on this topic.

2.1 Birds of a Feather

In a social network, nodes may have many properties, such as age, gender identity, ethnicity, sexual preference, location, topics of interests, and so on. Often, nodes that are connected to each other in a social network tend to be similar in their features: for example, relatives may live near each other, and friends may have similar interests. The technical name of this property is *assortativity*. Figure 2.1 illustrates assortativity based on a node feature represented by color. A more striking, real-world example from Twitter is shown in Figure 0.3. Because of assortativity, we are able to make predictions about a person's qualities by inspecting their neighbors. For instance, as we have seen in Section 0.1, researchers found that it is possible to ascertain with reasonable accuracy a Facebook user's sexual orientation and a Twitter user's political preference, even when these features are not present in their profile, by analyzing their circles of friends.

Multiple factors may be responsible for the presence of assortativity in social networks. One possibility is that if people are similar in some way, they are more likely to *select* each other and become connected. This property is captured by a popular proverb: "birds of a feather flock together." Its technical name is *homophily*. Examples include people living in geographic proximity, or who practice the same sport or hobby — these individuals are likely to meet and become friends. Dating apps leverage this kind of homophily by recommending matches based on shared personality traits. The converse mechanism is when people who are friends become more similar over time, through the process of *social influence*. Humans are social animals who tend to imitate each other since birth. Our ideas, opinions, and preferences are strongly affected by our social interactions. It is difficult

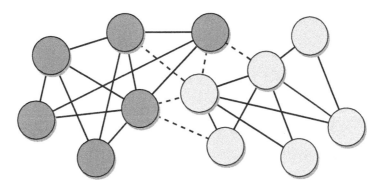

Fig. 2.1 Illustration of network assortativity. Nodes are more likely to be linked to other nodes of the same color than to nodes of different color. In particular, the majority of links for each node go to nodes of the same color, and the majority of links connect nodes of the same color. The few links connecting nodes of different colors are shown as dashed lines.

to separate the causes of assortativity — does similarity induce links, or do links induce similarity? Often these factors simultaneously shape our social connections and reinforce each other. We will return to this question in Chapter 7.

There is a dark side of homophily, too. On social media, it is exceedingly easy to connect with people who share our worldviews and unfriend or unfollow people with different opinions — all it takes is a tap of our finger. Furthermore, information can be shared and consumed in such a selective and efficient way as to influence our opinions very effectively. These mechanisms can lead to the segregation and polarization of our online communities, as is evident from Figure 0.3. When we are surrounded by people who reflect our own views, we are in an *echo chamber* — all the information and opinions to which we are exposed reflect our own and confirm or reinforce our ideas, rather than challenging them. This can be dangerous because it makes us vulnerable to manipulation by misinformation and social bots, as we will see in Chapter 4.

In NetworkX, we can calculate the assortativity of a network based on a given node attribute. There are two functions for cases when the attribute is categorical, such as gender, or numeric, such as age:

```
assort_a = nx.attribute_assortativity_coefficient(G, category)
assort_n = nx.numeric_assortativity_coefficient(G, quantity)
```

Assortativity is not exclusive of social networks; nodes in many types of networks have properties that may be similar among neighbors. For example, nodes in any network have the fundamental property of degree. Assortativity based on degree is called *degree assortativity* or *degree correlation*: this occurs when high-degree nodes tend to be connected to other high-degree nodes, while low-degree nodes tend to have other low-degree nodes as neighbors. Networks with this property are called *assortative*.

An example of an assortative network is shown in Figure 2.2(a): the hubs form a densely connected *core* while low-degree nodes are loosely attached to each other and/or to core

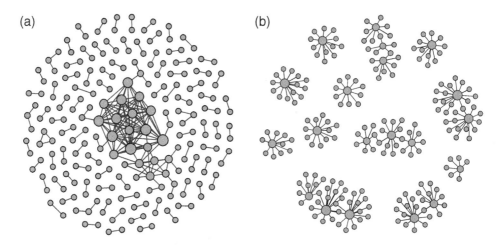

Fig. 2.2 Network degree assortativity illustrated by (a) an assortative network and (b) a disassortative network.

nodes. Therefore we say that assortative networks have a *core–periphery structure* (discussed in greater detail in Chapter 3). Social networks are often assortative. Networks where high-degree nodes tend to be connected to low-degree nodes and vice versa are called *disassortative*. An example is shown in Figure 2.2(b): hubs are situated at the center of star-like components. The Web, the Internet, food webs, and other biological networks tend to be disassortative.

There are two ways to measure the degree assortativity of a network, both based on measuring the *correlation* between degrees of neighbor nodes. We say that two variables are *positively (negatively) correlated* if larger values of one variable tend to correspond to larger (smaller) values of the other. Pearson's correlation coefficient is a common way to measure correlation; it takes values in $[-1, +1]$, with 0 meaning no correlation and ± 1 meaning perfect positive/negative correlation.

One measure of network assortativity is the *assortativity coefficient*, defined as the Pearson correlation between the degrees of pairs of linked nodes. Using NetworkX:

```
r = nx.degree_assortativity_coefficient(G)
```

When the assortativity coefficient is positive, the network is assortative, and when it is negative, the network is disassortative.

The second method is based on measuring the average degree of the neighbors of node i:

$$k_{nn}(i) = \frac{1}{k_i} \sum_j a_{ij} k_j, \tag{2.1}$$

where $a_{ij} = 1$ if i and j are neighbors, and 0 otherwise. We then define the *k-nearest-neighbors* function $\langle k_{nn}(k) \rangle$ for nodes of a given degree k as the average of $k_{nn}(i)$ across all nodes with degree k. If $\langle k_{nn}(k) \rangle$ is an increasing function of k, then high-degree nodes tend to be connected to high-degree nodes, therefore the network is assortative; if $\langle k_{nn}(k) \rangle$ decreases with k, the network is disassortative. Using NetworkX we can calculate the correlation between the degree and its associated neighbor connectivity:

```
import scipy.stats
knn_dict = nx.k_nearest_neighbors(G)
k, knn = list(knn_dict.keys()), list(knn_dict.values())
r, p_value = scipy.stats.pearsonr(k, knn)
```

Note that we need the `scipy` package for the Pearson correlation.

In Chapter 4 we will learn about a type of homophily based on information content that is very important for the Web.

2.2 Paths and Distances

If it is possible to go from a *source* node to a *target* node by traversing links in the network, we say that there is a *path* between the two nodes. The path is the sequence of links traversed. The number of links in a path is called the *path length*. There may be multiple paths between the same two nodes. These paths may have different lengths, and may or may not share some common links. In directed paths, we must comply with link directions. A *cycle* is a special path that can be traversed to go from one node back to itself. A *simple path* never goes through the same link more than once; in this book we focus only on simple paths. Finding paths was the earliest problem studied in network science (Box 2.1).

The concept of a path is the basis of the definition of *distance* among nodes in a network. The natural distance measure between two nodes is defined as the minimum number of links that must be traversed in a path connecting the two nodes. Such a path is called the *shortest path*, and its length is called the *shortest-path length*. There may be multiple shortest paths between two nodes; obviously they all must have the same length. In Section 2.5 we will see how to find the shortest path between two nodes. In some cases, such as transportation networks, one can imagine that a link is associated with a geographic distance between the adjacent nodes. In such cases we can redefine the path length as the *sum of the distances* associated with the links along the path; the length of a path from Berlin to Rome by way of Paris is the sum of the distances from Berlin to Paris and from Paris to Rome. You can think of an unweighted network as a special case in which all links have distance one.

Figure 2.3 illustrates the shortest paths between two nodes in different types of network, which depend on whether the network is directed and/or weighted. In the case of an undirected, unweighted network, the shortest path is just the one that minimizes the number of links traversed, and it is the same irrespective of the direction in which we move between

Box 2.1   The Seven Bridges of Königsberg

In 1736, Leonhard Euler used graph theory to solve a mathematical problem for the first time. The Prussian city of Königsberg was divided by the Pregel river into four land masses (North and South banks, Kneiphof and Lomse islands) connected by seven bridges. The problem was to devise a walk through the city that would cross each bridge once and only once. Euler formulated a generalized version of this problem as finding a path through a network where nodes and links represent the land masses and bridges, respectively, and each link is to be traversed exactly once.

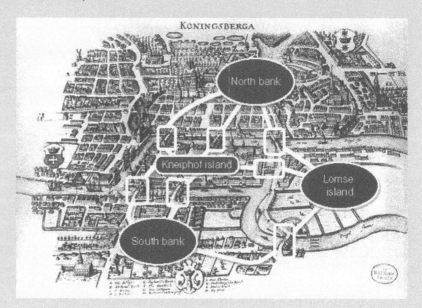

Euler proved that such a path (now called an *Eulerian path* in his honor) exists only if all nodes have even degree, except the source and the target. Nodes must have even degree because for each incoming link to arrive at a node, there must be an outgoing link to depart from the node. The source and the target, if distinct, must have odd degree because when the path starts (ends) it does not "cross" the node. If they coincide (*Eulerian cycle*), there cannot be nodes with odd degree. Since all four nodes in the Königsberg network have odd degree, there was no Eulerian path in this case.

the nodes. Note that there are two paths between nodes **a** and **b**, but the one that passes through node **d** is longer by one link; the shortest path bypasses node **d** by following the link between nodes **e** and **f**. The shortest-path length is $\ell_{ab} = 4$. The directed, unweighted network case is different because directed paths must be consistent with the direction of the links along the path. Therefore there is only one path from source **a** to target **b**, and it goes through **d**. The shortest directed path length is $\ell_{ab} = 5$. Beware that in a directed network there may be no paths between some pairs of nodes. For instance, if a node has only incoming links there are no paths having that node as the source. In the example of

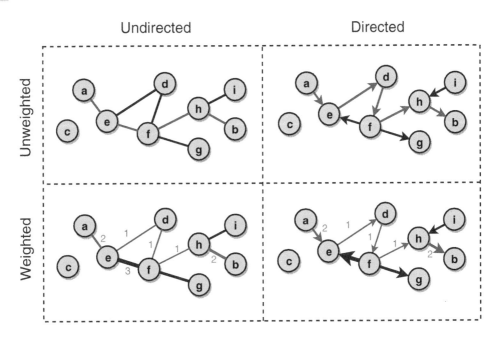

Fig. 2.3   Shortest paths in undirected, directed, unweighted, and weighted networks. Link weights represent distances and are shown in red. In each case the shortest path between nodes **a** and **b**, or from **a** to **b** in directed cases, is highlighted in blue. There is no path between node **c** and any other node. In directed networks, the shortest path must be consistent with the direction of the links along the path; there is no directed path from **b** to **a**.

Figure 2.3, there are no paths from **g** to any other node. Similarly, there are no paths to nodes that have only outgoing links, such as **a**.

The undirected, weighted network in Figure 2.3 shows what happens when we use link distances. In this case the shortest path between **a** and **b** goes through **d**: it has an extra link, but the sum of the distances between **e** and **f** through **d** is $1 + 1 = 2$, which is less than the distance 3 associated with the link (**e**, **f**). The directed, weighted case is straightforward: the shortest path is obtained by minimizing the sum of the distances along the path, while respecting the directions of the links. In both weighted network examples, the shortest-path length is $\ell_{ab} = 7$.

In many networks, link weights express a measure of similarity or intensity of interaction between two connected nodes. We may then be interested in finding paths with large weights. A common approach is to transform the weights into distances by taking the inverse (one divided by the weight), so that a large weight corresponds to a short distance. Then the problem becomes equivalent to finding short-distance paths.

By using the shortest-path length as a measure of distance among nodes, it is possible to define aggregate distance measures for an entire network: the *average shortest-path length* (or simply *average path length*) is obtained by averaging the shortest-path lengths across all pairs of nodes. The *diameter* of the network is instead the maximum shortest-path length across all pairs of nodes (i.e. the length of the longest shortest path in the network). The name is inspired by geometry, where the diameter is the longest distance between any two points on a circle.

Formally we define the *average path length* of an undirected, unweighted network as

$$\langle \ell \rangle = \frac{\sum_{i,j} \ell_{ij}}{\binom{N}{2}} = \frac{2\sum_{i,j} \ell_{ij}}{N(N-1)}, \tag{2.2}$$

where ℓ_{ij} is the shortest-path length between nodes i and j, and N is the number of nodes. The sum is over all pairs of nodes, and we divide by the number of pairs to compute the average. In the directed network case the definition is analogous, but the distance ℓ_{ij} is based on the shortest directed path between i and j, and each pair of nodes is considered twice for paths in both directions:

$$\langle \ell \rangle = \frac{\sum_{i,j} \ell_{ij}}{N(N-1)}. \tag{2.3}$$

The weighted cases are similar, with ℓ_{ij} defined based on link distances. The *diameter* of a network is

$$\ell_{max} = \max_{i,j} \ell_{ij}. \tag{2.4}$$

The definitions of average path length and diameter assume that the shortest-path length is defined for each pair of nodes. If there are any pairs without a path, then the average path length and diameter are not defined. For example, the networks in Figure 2.3 have no path between the singleton node **c** and any other node. We can think of such missing paths as paths with infinite distance. There are a few ways to deal with these cases.

If one wishes to define the average path length in a network where some of the paths do not exist, the following formula can be used for undirected networks:

$$\langle \ell \rangle = \left(\frac{\sum_{i,j} \frac{1}{\ell_{ij}}}{\binom{N}{2}} \right)^{-1}. \tag{2.5}$$

Note that if there is no path between i and j, $\ell_{ij} = \infty$ and therefore $1/\ell_{ij} = 0$ is defined. The same trick can be used for directed networks.

In Section 2.3 we show a couple of different ways to calculate the network distance and diameter when some paths are missing.

Both average path length and diameter can be used to describe the typical distance of a network. In this book we use the former. Although by definition the average cannot exceed the maximum, the two terms are sometimes used interchangeably because the two quantities behave similarly as the network size grows.

NetworkX has functions to determine the existence of paths, find shortest paths, and measure the length of a path or the average path length of a network. In the case of the undirected, unweighted network in Figure 2.3:

```
nx.has_path(G, 'a', 'c')              # False
nx.has_path(G, 'a', 'b')              # True
nx.shortest_path(G, 'a', 'b')         # ['a','e','f',h','b']
nx.shortest_path_length(G,'a','b')    # 4
nx.shortest_path(G, 'a')              # dictionary
nx.shortest_path_length(G, 'a')       # dictionary
nx.shortest_path(G)                   # all pairs
nx.shortest_path_length(G)            # all pairs
nx.average_shortest_path_length(G)    # error
G.remove_node('c')                    # make G connected
nx.average_shortest_path_length(G)    # now okay
```

When only the source node is specified, we obtain a dictionary with all shortest paths, or all shortest-path lengths, from the source. When neither source nor target are given, we get an object with the shortest paths for all pairs of nodes.

For directed networks, the functions are the same but they account properly for link directions. In the case of the directed, unweighted network in Figure 2.3:

```
nx.has_path(D, 'b', 'a')      # False
nx.has_path(D, 'a', 'b')      # True
nx.shortest_path(D, 'a', 'b') # ['a','e','d','f','h','b']
```

For weighted networks, we can store distances associated with links as weight attributes. We can then tell NetworkX to interpret the weights as distances when computing path lengths. In the case of the weighted, undirected network in Figure 2.3:

```
nx.shortest_path_length(W, 'a', 'b')           # 4
nx.shortest_path_length(W, 'a', 'b', 'weight') # 7
```

2.3 Connectedness and Components

To relate network structure and function, it is useful to consider the *connectedness* of a network. The connectedness defines many properties of a network's physical structure. For example, in Chapter 3 it will allow us to study the robustness of a network.

Recall from Chapter 1 that the number of links in a network is bound by the number of nodes. This is an upper bound; there is no lower bound, as a network might have no links at all, uninteresting as that may be. As we shall see in Chapter 5, the higher the density, the greater the chances that the network is *connected* (i.e. that you can reach any node from any other node by following a path along links and intermediate nodes). The fewer the links and the lower the density, the higher the chances that the network is disconnected, so that there are multiple nodes or groups of nodes that are not reachable from each other.

NetworkX has algorithms to determine whether a network is connected. For example, the networks in Figure 1.2 are all connected:

```
K4 = nx.complete_graph(4)
nx.is_connected(K4)          # True
C = nx.cycle_graph(4)
nx.is_connected(C)           # True
P = nx.path_graph(5)
nx.is_connected(P)           # True
S = nx.star_graph(6)
nx.is_connected(S)           # True
```

If a network is not connected, we say that it is *disconnected*; it is composed of more than one *connected components* or simply *components*. A component is a subnetwork containing one or more nodes, such that there is a path connecting any pair of these nodes, but there is no path connecting them to other components. The largest connected component in many real networks includes a substantial portion of the network and is called the *giant component*. In a connected network, the giant component coincides with the entire graph.

Figure 2.4 illustrates the components of a network, which are defined differently based on undirected and directed paths in undirected and directed networks, respectively. In the undirected case, the example in the figure contains three components. Note that by definition, a singleton belongs to its own component because it is not connected to any other nodes. In the directed case, things are a bit more complicated because we have to pay attention to link directions when determining whether a node can be reached from another. We can of course ignore the link directions and treat the links as if they were undirected. In this case we refer to the components as *weakly connected*. The directed network in Figure 2.4 has three weakly connected components. However, not all nodes in a weakly connected component may be reachable from each other following *directed* paths.

In a *strongly connected* component, there is at least one directed path between every pair of nodes, in both directions. In Figure 2.4, the largest strongly connected component

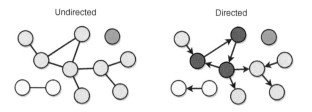

Undirected   Directed

Fig. 2.4 Connected components. Colors indicate different components. In the undirected network example we observe three components, one of which is a singleton. The light blue nodes make up the giant component. In the directed example, we observe three *weakly* connected components. The largest weakly connected component contains nodes of different shades of blue; the dark blue nodes make up the largest *strongly* connected component.

contains three nodes; every other node belongs to its own strongly connected component. Note that in a strongly connected network or component, there is at least one directed cycle from each node. To see why, consider any two nodes **a** and **b** in a strongly connected network. Since there must be a directed path from **a** to **b** and one from **b** to **a**, a cycle can be constructed by combining the two.

Having defined connected components, let us return to the issue of measuring network distance when a network is disconnected. One way is to consider only the nodes in the giant component. Another approach is to average the distance only across pairs of nodes in the same component, but considering all components. To calculate the diameter of a disconnected network, one can calculate the diameter of each component and then take the maximum.

We can identify the set of nodes from which one can reach a strongly connected component S, but that cannot be reached from S — if they could, they would be part of S. This set is called the *in-component* of S. Similarly, we define the *out-component* of S as the set of nodes that can be reached from S but from which one cannot reach S.

We say that a directed network is *strongly connected* if it is a single strongly connected component. A directed network is *weakly connected* if it is a single weakly connected component.

NetworkX provides functions to identify the connected components of a network. Assume G and D are the undirected and directed networks in Figure 2.4, respectively:

```
nx.is_connected(G)                        # False
comps = sorted(nx.connected_components(G),
               key=len, reverse=True)
nodes_in_giant_comp = comps[0]
GC = nx.subgraph(G, nodes_in_giant_comp)
nx.is_connected(GC)                       # True
nx.is_strongly_connected(D)               # False
nx.is_weakly_connected(D)                 # False
list(nx.weakly_connected_components(D))
list(nx.strongly_connected_components(D)) # lots of
                                          # singletons
```

In this example, we make use of the built-in function `sorted()` to list and sort the output of the `connected_components()` function. We specify `key=len` in order to sort by the component sizes, and `reverse=True` to output in descending order. Then the first element is the giant component.

2.4 Trees

Let us introduce a special class of undirected, connected networks such that the deletion of any one link will disconnect the network into two components. Such graphs are called *trees*.

The number of links in a tree is $L = N - 1$. To convince yourself that this is the case, start with a network with $N = 2$ nodes, which needs $L = 1$ link to be connected. Then, as we add one node at a time, we must add a link to connect the new node to some existing node. So the number of links is always equal to the number of nodes minus one. Removing any link will disconnect at least one node.

Trees have other interesting properties. They have no cycles. We can prove that trees cannot have cycles by contradiction: if a tree had a cycle, we could remove at least one link of the cycle without disconnecting it. Therefore the network would not be a tree — a contradiction. Because there are no cycles, given any pair of nodes, there is only a single path connecting them.

Trees are *hierarchical*. You can pick any node in a tree and call it a *root*. Each node in a tree is connected to a parent node (toward the root) and to one or more children nodes (away from the root). The exceptions are the root, which has no parent, and the so-called *leaves* of the tree, which have no children. The hierarchical structure of trees is illustrated in Figure 2.5.

NetworkX has algorithms to determine whether a network is a tree. For example, a complete network with more than two nodes has cycles and therefore is not a tree. The star and path networks in Figure 1.2 are examples of trees:

```
K4 = nx.complete_graph(4)
nx.is_tree(K4)              # False
C = nx.cycle_graph(4)
nx.is_tree(C)              # False
P = nx.path_graph(5)
nx.is_tree(P)              # True
S = nx.star_graph(6)
nx.is_tree(S)              # True
```

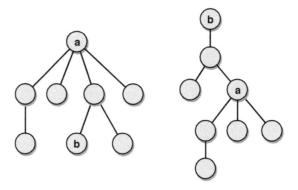

Fig. 2.5 Hierarchical structure of trees. The same tree is depicted with two different layouts, respectively with nodes **a** and **b** taken as roots and positioned at the top. Each node has its parent above (the root has no parent) and its children below (the leaves are at the bottom and have no children).

2.5 Finding Shortest Paths

In Section 2.2 we discussed shortest paths. But how to find the shortest path between two nodes in practice? To do this, it is necessary to map and navigate through the whole network. This is done by NetworkX and other network analysis tools. As we will see in Chapter 4, it is also done by search engines through *Web crawlers*, computer programs that automatically surf the Web finding and storing new pages.

The algorithm, or procedure for navigating through a network starting from a *source* node and finding the shortest path between the source and every other node in the network is called *breadth-first search*. The idea is that we visit the entire "breadth" of the network, within some distance from the source, before we move to a greater "depth," farther away from the source. This process is illustrated in Figure 2.6 for a simple undirected network: starting from some source node, we visit its neighbors (layer 1) and set the distance of those vertices from the source to one. We then visit the neighbors of the nodes in layer 1, except the nodes already explored (layer 2), and set their distance to two. Then we visit the neighbors of the layer 2 nodes if not previously visited (layer 3), setting their distance to three; and so on. Figure 2.6 shows that each layer contains all nodes having equal distance from the source. If the network is connected, all nodes are reached and assigned a distance from the source. The procedure is analogous for directed networks such as the Web, except that we only reach nodes through directed paths from the source.

To find the shortest paths from the source to the other nodes, the breadth-first search algorithm builds a directed *shortest-path tree*, containing the same nodes as the original network but only a subset of the links. The tree maps the shortest paths between its root (the source node) and all other nodes. The algorithm is illustrated in Figure 2.7 for a directed network; for implementation details, see Box 2.2.

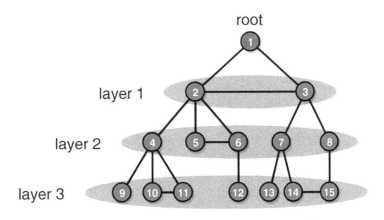

Fig. 2.6 Breadth-first search. In this instance, node **1** is selected as the source. First, we visit the neighbors of **1**, which are **2** and **3**. This is layer 1, including all nodes one step away from the source. Then we move to their neighbors **4, 5, 6, 7, 8**, which are two steps away from the source (layer 2). Finally, we reach nodes **9, 10, 11, 12, 13, 14, 15**, at distance three from the source (layer 3).

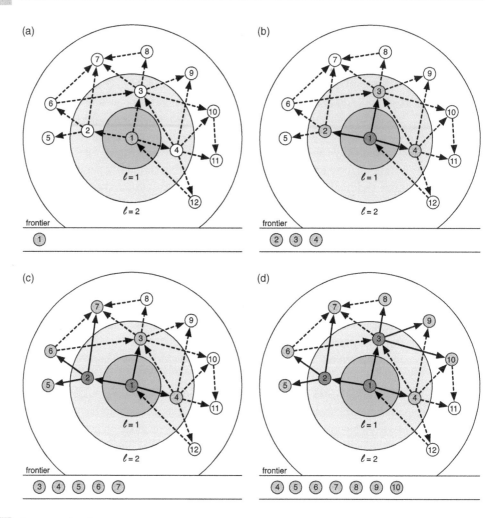

Fig. 2.7 Illustration of the breadth-first search algorithm to traverse a directed network and find short paths from a source node. Nodes are colored in light gray when they are added into the frontier queue, and dark gray when they are dequeued. Links turn from dashed to solid lines when they are added to the shortest-path tree. (a) The frontier is initialized with the source node **1**. (b) Node **1** is dequeued and its successors **2**, **3**, and **4** are queued into the frontier. (c) Node **2** is dequeued and its successors **5**, **6**, and **7** are queued. (d) Node **3** is dequeued and its successors **8**, **9**, and **10** are queued. Node **7** is already in the frontier, therefore the link to it from node **3** is ignored. The breadth-first algorithm keeps track of visited nodes because their distances are set, so they are not queued into the frontier again. For example, when node **4** is visited, successor node **3** can be ignored because its distance from the source is already set to one. In the next step, all nodes at distance one from the source will have been visited, so that nodes at distance two can be visited.

After the breadth-first search algorithm has been executed, all nodes in the same connected component as the source node have been assigned a distance from the source. To find the shortest path from the source to any target node, we have to follow the links in the shortest-path tree backward from the target node through predecessors in the upper layers, until we arrive at the source. Recall that in a tree there is a single path to the root; each

Box 2.2   Breadth-First Search

Breadth-first search takes a source node as input. To implement the algorithm, each node must have an attribute used to store its distance from the source. In addition, we must keep a queue of nodes that we call the *frontier*. A queue is a first-in-first-out data structure: nodes are extracted (dequeued) in the order in which they are inserted (queued).

Initially, the source node s is queued into the frontier. Its distance is set to $\ell(s, s) = 0$, and for all other nodes the distance is set to a common unrealistic value, say -1. The network that will eventually become the shortest-path tree is initialized with no links.

In each iteration, we visit the next node i in the frontier. The node is dequeued. Then, for each successor j of i (or each neighbor if the network is undirected), we follow three steps unless j already has its distance set:

1. Queue j into the frontier.
2. Set the distances of j from the source to $\ell_{s,j} = \ell_{s,i} + 1$.
3. Add a directed link $(i \rightarrow j)$ to the shortest-path tree.

The procedure ends when the frontier is empty. If any nodes remain with unknown distance, they are not reachable from the source; they must be in a different connected component of the network.

node has a single predecessor. Then we have to reverse the path to obtain the shortest path from the source to the target. In an undirected network this is the same as the path from the target to the source, but in a directed network they may be different.

In the example of Figure 2.7, we can see the shortest-path tree as its links are drawn with solid lines. For example, say we are interested in the shortest path from node **1** to node **7**. Breadth-first search has set the length of this path to $\ell_{1,7} = 2$. To find the path, we go from **7** to its predecessor in the shortest-path tree, which is node **2**, and then to its predecessor, which is the root node **1**. Reversing this path we obtain the shortest path **1**→**2**→**7**. Note that this is not the only short path — the path **1**→**3**→**7** has the same length, but the algorithm only identifies *one* shortest path from the source. Note also that in this directed network, the shortest path from node **7** to node **1** is not the same; in fact there is no such path.

The breadth-first search algorithm finds the shortest paths from a single source to all other nodes in an unweighted network. Slightly more complicated algorithms also exist for shortest paths in weighted networks. If we wish to find the shortest path between every pair of nodes, we need to run the algorithm N times, once from each node as a source. This is computationally expensive. In fact, as an exercise, you should try to use the NetworkX method `shortest_path(G)` [or `shortest_path_length(G)`] on some of the networks in the book's GitHub repository[1] (see also Table 2.1). You will notice that it takes a painfully long time for large networks; even if short paths exist, they are not necessarily easy to find. Fortunately, as we will see in Section 7.4, networks often provide us with clues so that we can search for a target node efficiently by following heuristic rules.

[1] github.com/CambridgeUniversityPress/FirstCourseNetworkScience

Table 2.1 Average path length and clustering coefficient of various network examples. The networks are the same as in Table 1.1; their numbers of nodes and links are listed as well. Link weights are ignored. The average path length is measured only on the giant component; for directed networks we consider directed paths in the giant strongly connected component. To measure the clustering coefficient in directed networks, we ignore link directions

Network	Nodes (N)	Links (L)	Average path length $(\langle \ell \rangle)$	Clustering coefficient (C)
Facebook Northwestern Univ.	10,567	488,337	2.7	0.24
IMDB movies and stars	563,443	921,160	12.1	0
IMDB co-stars	252,999	1,015,187	6.8	0.67
Twitter US politics	18,470	48,365	5.6	0.03
Enron email	87,273	321,918	3.6	0.12
Wikipedia math	15,220	194,103	3.9	0.31
Internet routers	190,914	607,610	7.0	0.16
US air transportation	546	2,781	3.2	0.49
World air transportation	3,179	18,617	4.0	0.49
Yeast protein interactions	1,870	2,277	6.8	0.07
C. elegans brain	297	2,345	4.0	0.29
Everglades ecological food web	69	916	2.2	0.55

2.6 Social Distance

The average path length, defined in Section 2.2, characterizes how close or far we expect nodes to be in a network. Intuitively, in a grid-like network like road networks and power grids, paths can be long. Is this typical of many real-world networks? Let us start by considering a few social networks, in which this question has been explored extensively.

Coauthorship networks are a well-studied kind of social collaboration network because it is relatively easy to gather data about nodes and links. Nodes are scholars, and links can be mined from digital libraries. When we see a publication coauthored by two or more scholars, we can infer links between them in the network.

Paul Erdős was a famous mathematician who made critical contributions to network science, discussed in Chapter 5. (For more background about his life, see Box 5.1.) Mathematicians are fond of studying their distance in the coauthorship network from the particular node corresponding to Erdős. They call this distance their *Erdős number* (Box 2.3). Many mathematicians have a very small Erdős number. Figure 2.8 illustrates the network of collaborations involving Erdős and his over 500 coauthors. In reality, scholars are not just close to Erdős; they are close to everyone. This is typical of collaboration networks: there are short paths among all pairs of nodes. Pick any two scholars and they will not be very far from each other.

Box 2.3   The Erdős Number

Paul Erdős was one of the world's greatest mathematicians. He also stands out among scientists due to his amazing productivity and number of collaborators. Therefore Erdős plays an important role in the connectedness of the scientific collaboration network, in that one can go from many nodes of the graph to many others through him. This is so much so that a special measure has been defined in his honor: the *Erdős number*. Many scientists proudly display their Erdős number on their homepages and CVs. This number is simply defined as the length of the shortest path, in the coauthorship network, from a scholar to Paul Erdős. There is even an online tool to compute the Erdős number for mathematicians (www.ams.org/mathscinet/collaborationDistance.html). For example, Erdős was a collaborator of Fan Chung, who coauthored a report with Alex Vespignani, who has been a coauthor of two of the authors of this book, who therefore have an Erdős number of three. Because of the huge number of Paul Erdős's coauthors, the number of scholars with a small Erdős number is quite large.

It turns out that not only collaboration networks, but pretty much all social networks have very short paths among nodes. You are likely to know someone who knows someone who knows someone...and in a few steps you can get to anyone on the planet! For a demonstration from a more familiar domain, let us turn to the social network connecting movie stars. As we have seen in Chapter 0, nodes are actors and actresses, and two nodes are linked if they have co-starred in a movie. *Six Degrees of Kevin Bacon* is a fun game that originates from such a network. The game, illustrated in Figure 2.9, consists of finding the shortest path connecting an arbitrary actress or actor to Kevin Bacon in the co-star network. For example, a path of length $\ell = 2$ connects Marilyn Monroe to Kevin Bacon. You can play this game online at *The Oracle of Bacon* (oracleofbacon.org). The website pulls data to build the network from the Internet Movie Database (IMDB.com). While Kevin Bacon is often jokingly considered "the" hub of the star network, in reality he is not special; you can enter any pair of actors/actresses and the Oracle shows you the shortest path as a sequence of nodes (stars) and links (movies). Can you find two familiar stars separated by more than four links? Play this game and try!

The Erdős number and the Oracle of Bacon demonstrate that finding long paths in real-world networks is not easy. When we think about it, the concept of short social distance — that we are all only a few steps from each other in the social network — is a familiar one. How many times have you met someone and then been surprised to discover a common friend? The low expectation of running into a friend of a friend is rooted in our intuition of how small our circle of acquaintances is, compared to an entire population. Yet this sort of thing happens often enough, prompting us to exclaim "what a small world!" A *small world* is the popular notion that social distances are short, on average. Consequently, the number of friends of friends out there is much, much larger than we think, and finding short paths in the social network is not so strange after all.

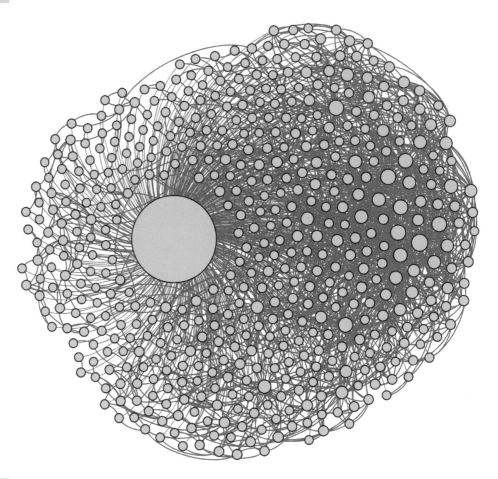

Fig. 2.8 Ego network of Paul Erdős (the large node at the center) within the coauthorship network. Ego networks are defined in Section 1.4.

2.7 Six Degrees of Separation

The name of the game *Six Degrees of Kevin Bacon* is inspired by the concept of *six degrees of separation*. The idea is the same as that of a small world: any two people in the world are connected by a short chain of acquaintances. In other words, social networks have a short diameter and an even shorter average path length. The number "six" in the expression originated from Hungarian author Frigyes Karinthy in the 1920s, and some credit goes to Italian inventor Guglielmo Marconi as well for coming up with the same idea 20 years before, in the early 1900s. However, what made the "six degrees" expression famous was an experiment conducted by psychologist Stanley Milgram[2] in the 1960s, which provided the first empirical evidence of small worlds.

---

[2] Milgram is famous for another, very controversial experiment, in which subjects were instructed to inflict pain on other people. The goal was to test the degree to which a human being is capable of immoral acts as a result of pressure from authority.

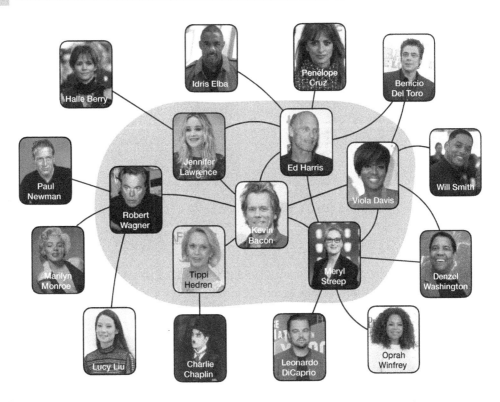

Fig. 2.9 Illustration of the *Six Degrees of Kevin Bacon* game. A few of the nodes connected to Kevin Bacon in the co-star network are shown in the shaded area, along with links among them. A small sample of the nodes at distance $\ell = 2$ is also included. Photo credit: Getty Images.

Milgram wanted to measure the social distance between strangers. He therefore asked 160 subjects in Nebraska and Kansas to forward a letter to an acquaintance, with instructions that the letter should eventually reach a target person in Massachusetts. Each recipient was supposed to forward the letter to someone known, who was likely to know the target. Only 42 of the letters (26%) reached the target. In those cases, however, the path lengths were surprisingly short, ranging between 3 and 12 steps. Figure 2.10 illustrates a typical path of 4 steps. The average path length was a little more than 6 steps, which would eventually inspire a play titled "six degrees of separation" that ultimately popularized the small-world notion. Migram's experiment was replicated in 2003 using emails to recruit a larger number of subjects. There were 18 targets in 13 countries. Of the more than 24,000 chains started, only 384 were completed, with an average length of 4 steps. When accounting for the many broken chains, the authors estimated a median path length of 5–7 steps, in agreement with Milgram's "six degrees." Even more recently, in 2011, researchers at Facebook and the University of Milan examined all 721 million Facebook users who were active at that time (more than 10% of the global population), with 69 billion friendships among them, and found that the average path length was 4.74 steps.

So far we have been referring to the paths we find when playing a game like the *Six Degrees of Kevin Bacon*, or those reported in the studies by Milgram and other researchers,

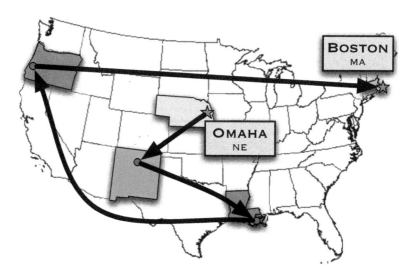

Fig. 2.10   The path followed by one of the letters in Milgram's experiment. The source subject in Omaha, NE sent the letter to an acquaintance in Santa Fe, NM. From there the letter was forwarded to people in New Orleans, LA and Eugene, OR before reaching the target in Boston, MA.

as "short." But when can we call a path *short*? Compared to what? Would we call a path with 6 steps *short* in a network with only 10 nodes? Clearly we must define what we mean by *short* paths more precisely, and the definition must be relative to the size of the network. In fact, it makes more sense to observe the relationship between the average path length $\langle \ell \rangle$ and the network size N when we consider networks (or subnetworks) of different sizes. We say that the average path length is *short* when it grows very slowly with the size of the network.

We can express slow growth mathematically by saying that the average path length scales *logarithmically* with the size of the network:

$$\langle \ell \rangle \sim \log N.$$

The logarithm of a in base b, $\log_b a$, is the exponent c such that $b^c = a$. Base $b = 10$ is commonly used; $\log_{10} 10 = 1$ because $10^1 = 10$, $\log_{10} 100 = 2$ because $10^2 = 100$, $\log_{10} 1000 = 3$, etc. The logarithm is therefore a function that grows very slowly.

What this means is that the network could have tens of millions of nodes, and yet its average path length would be in the single digits. Furthermore, the network could multiply many times in size while the average path length would only get a few steps longer.

Short paths that obey this kind of relationship are found across social networks, including academic collaborations, actor networks, networks of high-school friends, and online social networks such as Facebook. Short social distances can be useful, say when we are looking for a job. But short paths are not an exclusive feature of social networks. In fact, searching for paths is something we do routinely in all kinds of networks, for example when we book a long flight and try to minimize the number of intermediate stops. Finding network paths can be fun, too. *Wikiracing* is a hypertext search game designed to work with Wikipedia. A player must navigate from a source article to a target article, both randomly selected, solely by clicking links within each article. The goal is to reach the target in the fewest clicks (i.e. to find a network path with few links). There are variations for teams and with a race against the clock. You can play several versions of this game online, such as *The Wiki Game* (thewikigame.com). You will be amazed at how quickly you can reach any target with a bit of practice. This tells us that Wikipedia has short paths. The same is true for the Web, as we will see in Chapter 4.

As it turns out, short paths are a ubiquitous feature in almost all real-world networks; grid-like networks are among the few exceptions. Table 2.1 reports the average path length of various networks.[3] In all of these examples, the average path length is only a few steps. In the case of the movies and stars network, the paths appear to be longer. However, keep in mind that this is a bipartite network in which a link connects a movie and an actor/actress. If we consider the co-star network, in which two stars are connected if they acted together (as in Figure 2.9), links are associated with movies; in this case the average path length is roughly cut in half. In Chapter 5 we will find short paths even in the simplest of networks, where links are assigned at random.

2.8 Friend of a Friend

In a social network, if Alice and Bob are both friends of Charlie's, they are also likely to be friends of each other. In other words, there is a good chance that a friend of my friend is also my friend. This translates into the presence of many *triangles* in the network. As illustrated in Figure 2.11(a), a *triangle* is a triad (set of three nodes) where each pair of nodes is connected. The connectivity among the neighbors of the nodes is an important feature of the local structure of the network because it captures how tightly knit, or *clustered*, the nodes are.

The *clustering coefficient* of a node is the *fraction of pairs of the node's neighbors that are connected to each other*. This is the same as the ratio between the number of triangles that include the node, and the maximum number of triangles in which the node *could* participate.

The *clustering coefficient* of node i is formally defined as

$$C(i) = \frac{\tau(i)}{\tau_{max}(i)} = \frac{\tau(i)}{\binom{k_i}{2}} = \frac{2\tau(i)}{k_i(k_i - 1)}, \tag{2.6}$$

---

[3] Datasets for these networks are available in the book's GitHub repository: github.com/Cambridge UniversityPress/FirstCourseNetworkScience

(a)   (b)

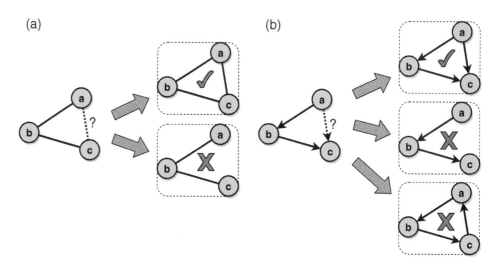

Fig. 2.11 Triads and triangles. (a) In an undirected network, node **b** has neighbors **a** and **c**. They may or may not form a triangle, depending on whether or not **a** and **c** are connected to each other. (b) In a directed network, node **a** links to **b** and node **b** links to **c**. A shortcut link from **a** to **c** would form a directed triangle.

where $\tau(i)$ is the number of triangles involving i. The maximum possible number of triangles for i is the number of pairs formed by its k_i neighbors. Note that $C(i)$ is only defined if the degree $k_i > 1$ due to the terms k_i and $k_i - 1$ in the denominator: a node must have at least two neighbors for any triangle to be possible.

The clustering coefficient of the entire network is the average of the clustering coefficients of its nodes:

$$C = \frac{\sum_{i:k_i>1} C(i)}{N_{k>1}}. \tag{2.7}$$

Nodes with degree $k < 2$ are excluded when calculating the average clustering coefficient.

Figure 2.12 illustrates how to calculate the clustering coefficient for a few nodes in a network. Node **a** has two neighbors **f** and **g** that are connected to each other, forming a triangle. Therefore its clustering coefficient is $C(a) = 1/1 = 1$. Node **b** has four neighbors. Only two of the six pairs of neighbors are connected: (**e**, **c**) and (**c**, **g**). Therefore, $C(b) = 2/6 = 1/3$. Node **c** has three neighbors that form two triangles via links (**e**, **b**) and (**b**, **g**). The third possible triangle is not realized because the link (**e**, **g**) is missing. Therefore, $C(c) = 2/3$. Finally, node **d** has a single neighbor **e**, therefore $C(d)$ is undefined.

Our definition of a clustering coefficient only applies to undirected networks, because we have only defined undirected triangles. We could extend the definition to directed networks, but it depends on the kinds of triangles that are relevant to a specific case. On Twitter, for example, we might be interested in triangles that shortcut paths along which information

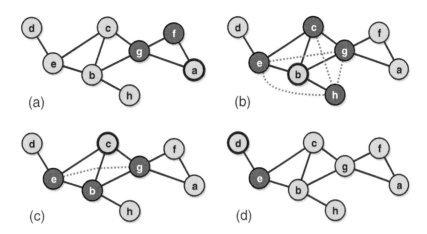

Fig. 2.12 Examples of clustering coefficient. (a) Node **a** has two neighbors **f** and **g** that are connected, forming a triangle. (b) Node **b** has four neighbors **c**, **e**, **g**, and **h**. Two of the six pairs of neighbors are connected, forming two out of six possible triangles. The missing triangle connections are shown by dotted gray lines. (c) Node **c** has three neighbors **e**, **b**, and **g** forming two out of three possible triangles. (d) Node **d** has a single neighbor **e**, therefore there are no possible triangles and the clustering coefficient is undefined.

travels. Consider the scenario in Figure 2.11(b): if **a** follows **b** and **b** follows **c**, then **a** might be interested in following **c** in order to access **c**'s posts directly rather than through **b**'s retweets. In such a scenario, we may want to only count directed triangles that encode these kinds of shortcuts. In this book we only deal with the clustering coefficient in undirected networks; for directed networks we can simply ignore the direction of the links and treat them as if they were undirected when calculating the clustering coefficient.

By averaging the clustering coefficient across the nodes, we can calculate a clustering coefficient for an entire network. A low clustering coefficient (near zero) means that the network has few triangles, while a high clustering coefficient (near one) means that the network has many triangles. Social networks have a large clustering coefficient; a significant portion of all possible triangles are present. For example, coauthorship networks tend to have clustering coefficient above 0.5. A simple mechanism explains the abundance of triangles in social networks: we meet people through shared contacts, thus closing triangles. This mechanism, called *triadic closure*, is discussed further in Chapter 5. Online social networks make suggestions based on triadic closure. For example, Facebook recommends "people you may know" based on common friends, and Twitter recommends accounts followed by your friends (whose accounts you follow). These recommendations result in high clustering.

Table 2.1 reports the clustering coefficient of various networks. We observe high clustering in many, but not all cases. The movies and stars network has $C = 0$. This is because the network is bipartite, therefore there can be no triangles; triangles would require links between pairs of movies or stars, which are not present in the bipartite network. If we instead examine the social network of co-stars, we find a high clustering coefficient. The Twitter retweet network also has a low $C = 0.03$. To understand why, consider that if Bob retweets Alice and Charlie retweets Bob, Twitter links both Bob and Charlie to the original author, Alice. Therefore each retweet cascade tree looks like a star. The only triangles stem from users participating in multiple stars.

NetworkX has functions to count triangles and calculate the clustering coefficient for nodes and networks. Currently NetworkX sets the clustering coefficient to zero for nodes with degree below two and includes those nodes in the average calculation.

```
nx.triangles(G)          # dict node -> no. triangles
nx.clustering(G, node)   # clustering coefficient of node
nx.clustering(G)         # dict node -> clustering coeff.
nx.average_clustering(G) # network's clustering coeff.
```

2.9 Summary

In this chapter we have learned about several features of networks: assortativity, connectedness, short paths, and clustering.

1. Assortativity is the correlation between the likelihood that two nodes are connected and their similarity. Similarity can be measured based on degree, content, location, topical interests, or any other node property. Assortativity in social networks can be due to homophily, the tendency of similar people to be connected; or to social influence, the tendency of connected people to be similar.
2. Paths are sequences of links connecting nodes in a network. The natural distance measure between two nodes is defined as the number of links traversed by the shortest connecting path. The simplest way to find a short path is the breadth-first search algorithm. The concepts of paths and distances can be extended to take into consideration link directions and weights.
3. A tree is a connected undirected network with as few links as possible. Trees have no cycles.
4. Connected components are subnetworks such that there exists a path between any two nodes in the same component, but not between two nodes in different components. In directed networks we distinguish between strongly and weakly connected components based on whether or not paths respect link directions.
5. The average path length of a network is found by averaging the shortest-path lengths across all pairs of nodes in a connected network. If a network is not connected, usually only pairs of nodes in the same component are considered.
6. Most real networks have very short paths on average. This is known as the small-world property. The popular notion that social networks have six degrees of separation originated from Milgram's experiment.
7. The local clustering of a network is induced by the presence of triangles, or connected triads. For a node, the clustering coefficient measures the fraction of triangles out of

the maximum possible number. For an entire network we can average the clustering coefficient across nodes. Social networks have high clustering due to friend-of-a-friend triangles.

2.10 Further Reading

The word "homophily" originates from the Greek "homós" (same) and "philia" (friendship). The concept was formulated by Lazarsfeld *et al.* (1954) and the presence of various forms of homophily has been observed in many studies of social networks (McPherson *et al.*, 2001). Aiello *et al.* (2012) found that users with similar interests are more likely to be friends in various online social media platforms, and that similarity among users based on their profile metadata is predictive of social links. The *k*-nearest-neighbors connectivity and assortativity coefficient were introduced by Pastor-Satorras *et al.* (2001) and Newman (2002), respectively.

Researchers are increasingly studying the negative consequences of homophily. Exposure to news and information through the filter of like-minded individuals in online social networks may facilitate the emergence of clustered communities in which our attention is focused toward information that we are already likely to know or agree with. These so-called "echo chambers" (Sunstein, 2001) and "filter bubbles" have been claimed to be pathological consequences of social media recommendation algorithms (Pariser, 2011) and to lead to polarization (Conover *et al.*, 2011b) and viral misinformation (Lazer *et al.*, 2018).

Algorithms for finding shortest paths and connected components in networks have a complicated history. The invention of breadth-first search is attributed to Zuse and Burke in a rejected 1945 Ph.D. thesis, and independently to Moore (1959). There are two famous algorithms for finding shortest paths in weighted networks: one by Dijkstra (1959) and the Bellman–Ford algorithm, published independently by Shimbel (1955), Ford Jr. (1956), Moore (1959), and Bellman (1958).

Milgram's experiment (Travers and Milgram, 1969) was repeated by Dodds *et al.* (2003) using emails. Backstrom *et al.* (2012) found that the average shortest-path length on the network of Facebook friends is lower than five. Newman (2001) first studied the structure of scientific collaboration networks.

An accessible introduction to networks and their small-world and clustered structure is offered by Watts (2004). The existence of triangles in networks is also referred to as *transitivity* (Holland and Leinhardt, 1971). An early definition of the network clustering coefficient was formulated by Luce and Perry (1949), while the local definition used in this book is due to Watts and Strogatz (1998).

The concept of triadic closure was introduced in a seminal paper by Granovetter (1973) and is discussed in Chapter 5. Studying data from a social media platform, Weng *et al.* (2013a) confirmed that triadic closure has a strong effect on link formation, but also found that shortcuts based on traffic are another key factor in explaining new links.

Exercises

2.1   Go through the Chapter 2 Tutorial on the book's GitHub repository.[4]

2.2   Recall that unless otherwise specified, the length of a path is the number of links contained therein. Given two nodes in an arbitrary undirected, connected graph, there must exist some shortest path between them. True or False: There may exist multiple such shortest paths.

2.3   True or False: Given any two nodes in an (undirected) tree, there exists exactly one path between those two nodes.

2.4   Consider an undirected, connected network with N nodes. What is the minimum number of links the network can have? If we do not require the network to be connected, does that minimum number of links change?

2.5   Recall that a tree of N nodes contains $N - 1$ links. True or False: Any connected, undirected network of N nodes and $N - 1$ links must be a tree.

2.6   True or False: Any undirected network of N nodes with at least N links must contain a cycle.

2.7   True or False: Any directed network of N nodes with at least N links must contain a cycle.

2.8   Consider the network defined by the adjacency matrix in Eq. (1.11). Are there any cycles in this network? Is it strongly connected? Weakly connected?

2.9   Consider the unweighted, undirected version of the network defined by the adjacency matrix in Eq. (1.11). Is this network a tree?

2.10   Consider the unweighted, undirected version of the network defined by the adjacency matrix in Eq. (1.11). What is this network's diameter?

2.11   If you convert a weakly connected directed network to an undirected network, will the resulting network be connected? Explain why or why not.

2.12   Consider an arbitrary non-complete undirected network. Now add a single link. How has the number of nodes in this network's giant component changed as a result of this addition?
   a. It has strictly decreased
   b. It has decreased or stayed the same
   c. It has increased or stayed the same
   d. It has strictly increased

2.13   Consider the weighted directed network in Figure 2.13. Which of the following most accurately describes the connectedness of this network?

---

[4] github.com/CambridgeUniversityPress/FirstCourseNetworkScience

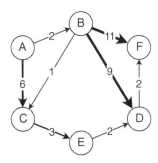

Fig. 2.13   A weighted, directed network. The numbers give the link weights.

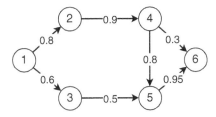

Fig. 2.14   A weighted, directed network. The numbers give the link weights.

       a. Strongly connected
       b. Weakly connected
       c. Disconnected
       d. None of the above

2.14 Consider the weighted directed network in Figure 2.13. What is the in-strength of node **D**? What is the out-strength of node **C**? (Recall the definitions from Chapter 1.)

2.15 How many nodes are in the largest strongly connected component of the network in Figure 2.13?

2.16 Consider the network in Figure 2.14. Which of the following most accurately describes the connectedness of this network?
       a. Strongly connected
       b. Weakly connected
       c. Disconnected
       d. None of the above

2.17 Link weights can represent anything about the relationship between the nodes: strength of the relationship, geographic distance, voltage flowing through a link cable, etc. When discussing path lengths on a weighted graph, one must first define how the weights are related to the distances. The length of a path between two nodes is then the sum of the distances of the links in that path. The simplest case occurs when the link weights represent the distance. Consider the network in Figure 2.14

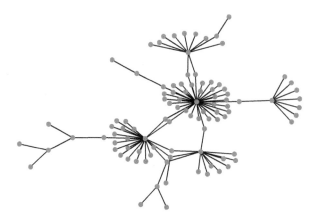

A small subnetwork of the *Drosophila melanogaster* (a.k.a. fruit fly) protein interaction network. Each node represents a protein that interacts with other proteins to perform the essential work of the cell. Experimental evidence has demonstrated that linked proteins form a molecular bond to accomplish some biological function.

and assume that the link weights represent distances. Using this distance metric, what is the shortest path between nodes 1 and 6?

2.18 A common way to define the distance between two nodes is the inverse (or reciprocal) of the link weight. Consider the network in Figure 2.14, and assume that the distance between two adjacent nodes is defined as the reciprocal of the link weight. Using this distance metric, what is the shortest path between nodes 1 and 6?

2.19 Consider the network in Figure 2.15. Which of the following is the best estimate of this network's diameter?
   a. 2
   b. 4
   c. 10
   d. 20

2.20 Consider the network in Figure 2.15. Which of the following is the best estimate for the average clustering coefficient of this graph?
   a. 0.05
   b. 0.5
   c. 0.75
   d. 0.95

2.21 Would a social network be likely to have the diameter and clustering coefficient of the graph in Figure 2.15?

2.22 Consider the network in Figure 2.16. Which of the following most accurately describes the connectedness of this network?
   a. Strongly connected
   b. Weakly connected

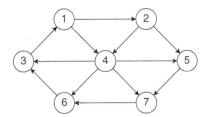

Fig. 2.16   A directed network.

        c. Disconnected

        d. None of the above

2.23 What is the diameter of the network in Figure 2.16?

2.24 Consider an undirected version of the network in Figure 2.16. What is the diameter of this network?

2.25 Consider any arbitrary directed graph D along with its undirected version G. True or False: If the average shortest-path length and diameter of the directed graph exist, they can be smaller than those of the undirected version.

2.26 Imagine that you were building a competitor of NetworkX. You have already written a method `shortest_path()` to compute the shortest path between two nodes, and now you want to write a function to compute the diameter of a network. Which of the following best describes how to go about doing this?

        a. First compute the shortest-path lengths between each pair of nodes. The diameter is the minimum of these values

        b. First compute the shortest-path lengths between each pair of nodes. The diameter is the average of these values

        c. First compute the shortest-path lengths between each pair of nodes. The diameter is the maximum of these values

        d. First compute the average length of all paths between each pair of nodes. The diameter is the minimum of these values

2.27 True or False: A network's diameter is always greater than or equal to its average path length.

2.28 What is the central idea behind the notion of "six degrees of separation"?

        a. Social networks have high clustering coefficients

        b. Social networks are sparse

        c. Social networks have many high-degree nodes

        d. Social networks have small average path length

2.29 The American Mathematical Society has a Web tool to find the *collaboration distance* between two mathematicians (see Box 2.3). Use this tool to calculate the Erdős number for a few mathematicians in your institution, or whom you know by fame.

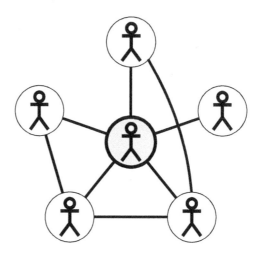

An ego network. The ego is highlighted in yellow.

2.30 Use *The Oracle of Bacon* (oracleofbacon.org) to measure the shortest-path distance in the co-star network among as many pairs of obscure actors and actresses as you can think of. Plot a histogram showing the distribution of the shortest-path lengths, and also estimate the average path length based on your sample. (If you are not familiar with histograms, they are defined in the next chapter.)

2.31 Play *The Wiki Game* (thewikigame.com) until you are able to complete a few rounds successfully. Report the average length (number of clicks) of the discovered paths.

2.32 What is the maximum clustering coefficient for a node in an arbitrary undirected graph?

2.33 What is the maximum clustering coefficient for a node in a tree?

2.34 Recall the definition of an ego network in Section 1.4. Consider the ego network in Figure 2.17: what is the clustering coefficient of the ego?

2.35 Consider the undirected network in Figure 2.4. Compute the shortest-path length for each pair of nodes in the giant component.

2.36 Consider the undirected network in Figure 2.4. Compute the clustering coefficient for each node such that it is defined.

2.37 Consider the network example in Figure 2.12. Compute the shortest-path length for each pair of nodes, and the average shortest-path length for the network.

2.38 Consider the network example in Figure 2.12. Compute the clustering coefficient for each node such that it is defined, and for the network.

2.39 If you use an online social network such as Facebook or LinkedIn, measure your clustering coefficient in the network. (*Hint 1*: If you use a social network with directed links, such as Twitter or Instagram, you can treat the links as undirected.) (*Hint 2*:

This might take a while; it's okay to make an estimate based on a small sample of your friends.)

2.40 Which of the following seemingly conflicting properties are true of social networks?
  a. Social networks have short paths, yet large diameter
  b. Social networks have small diameter, yet large average path length
  c. Social networks have many high-degree nodes, yet are disconnected
  d. Social networks are highly clustered, yet are not dense

2.41 The `socfb-Northwestern25` network in the book's GitHub repository is a snapshot of Northwestern University's Facebook network. The nodes are anonymous users and the links are friend relationships. Load this network into a NetworkX graph in order to answer the following questions. Be sure to use the proper graph class for an undirected, unweighted network.
  1. How many nodes and links are in this network?
  2. Which of the following best describes the connectedness of this network?
    a. Strongly connected
    b. Weakly connected
    c. Connected
    d. Disconnected
  3. We want to obtain some idea about the average length of paths in this network, but with large networks like this it is often too computationally expensive to calculate the shortest path between every pair of nodes. If we wanted to compute the shortest path between every pair of nodes in this network, how many shortest-path calculations would be required? In other words, how many pairs of nodes are there in this network? (*Hint*: Remember this network is undirected and we usually ignore self-loops, especially when computing paths.)
  4. To save time, let's try a sampling approach. You can obtain a random pair of nodes with

```
random.sample(G.nodes, 2)
```

  Since this sampling is done without replacement, it prevents you from picking the same node twice. Do this 1000 times and for each such pair of nodes, record the length of the shortest path between them. Take the mean of this sample to obtain an estimate for the average path length in this network. Report your estimate to one decimal place.
  5. Apply a slight modification to the above procedure to estimate the diameter of the network. Report the approximate diameter.
  6. What is the average clustering coefficient for this network? Answer to at least two decimal places.
  7. Is this network assortative or disassortative? Answer this question using the two methods shown in the text. Do the answers differ?

## 3

Hubs

hub: (*n.*) a center around which other things revolve or from which they radiate; a focus of activity, authority, commerce, transportation, etc.

If you have traveled on a plane, you have traversed an important network — the air transportation network. In Figure 0.7 we mapped the US air transportation network: the nodes represent airports and the links are direct flights between them. While most airports are rather small, a few major ones (e.g. Atlanta, Chicago, Denver) have daily flights to hundreds or even thousands of destinations. Similarly, in social communities there are individuals who are much more visible and influential than others; and on the Web there are some very popular sites, such as google.com, while most sites are unknown to most.

These examples illustrate a key feature of many networks: *heterogeneity*. Heterogeneous networks present a wide variability in the properties and roles of their elements — nodes and/or links. This reflects the diversity present in the complex systems described by networks. In air transportation networks, social networks, the Web, and many other networks, a clear source of heterogeneity is the degree of the nodes: a few nodes have many connections (Atlanta, Google, Obama), while most nodes have few.

The importance of a node or a link is estimated by computing its *centrality*. There are several ways to measure network centrality. In this chapter we introduce a few important centrality measures, for nodes in particular. As we discuss below, the degree is an important measure of centrality. High-degree nodes are called *hubs*. As it turns out, hubs are responsible for some striking properties that characterize a broad variety of networks.

3.1 Centrality Measures

3.1.1 Degree

In Chapter 1 we learned that the degree of a node is the number of neighbors of that node. In the example of the US airport network in Figure 0.7, the degree of a node (airport) is the number of other airports reachable from it via direct flights.

In a social network, the degree of a node (individual) is the number of social links connecting the node to others. For instance, in a coauthorship network such as the one depicted in Figure 2.8, the degree is the number of collaborators. High-degree nodes in social networks are people with many connections — whether because they are sociable, sought after, or simply eager to collaborate, these nodes seem to be important in some sense. Therefore, the degree is a very natural measure of centrality in social networks.

The *average degree* of a network indicates how connected the nodes are on average. As we shall see later (Section 3.2), the average degree may not be representative of the actual distribution of degree values. This is the case when the nodes have heterogeneous degrees, as in many real-world networks.

3.1.2 Closeness

Another way to measure the centrality of a node is by determining how "close" it is to the other nodes. This can be done by summing the distances from the node to all others. If the distances are short on average, their sum is a small number and we say that the node has high centrality. This leads to the definition of *closeness centrality*, which is simply the inverse of the sum of distances of a node from all others.

The closeness centrality of a node i is defined as

$$g_i = \frac{1}{\sum_{j \neq i} \ell_{ij}}, \tag{3.1}$$

where ℓ_{ij} is the distance from i to j and the sum runs over all the nodes of the network, except i itself. An alternative formulation is obtained by multiplying g_i by the constant $N - 1$, which is just the number of terms in the sum at the denominator:

$$\tilde{g}_i = (N - 1)g_i = \frac{N - 1}{\sum_{j \neq i} \ell_{ij}} = \frac{1}{\sum_{j \neq i} \ell_{ij}/(N - 1)}. \tag{3.2}$$

This way we discount the graph size and make the measure comparable across different networks. Since what matters is not the actual value of g_i but its ranking compared to the closeness centrality of the other nodes, the relative centrality of the nodes remains the same as by using Eq. (3.1), because the ranking is not altered if the values are multiplied by a constant. The expression $\sum_{j \neq i} \ell_{ij}/(N - 1)$ is the *average distance* from the focal node i to the rest of the network. So we find that closeness can be expressed equivalently as the inverse of the average distance.

NetworkX has a function to compute the closeness centrality:

```
nx.closeness_centrality(G, node)  # closeness centrality
                                  # of node
```

3.1.3 Betweenness

Many phenomena taking place in networks are based on diffusion processes (Chapter 7). Examples include the transmission of information across a social network, the traffic of goods through a port, and the spread of epidemics in the network of physical contacts between the individuals of a population. This has suggested a third notion of centrality, called *betweenness*: a node is the more central, the more often it is involved in these processes.

Naturally, betweenness centrality has a different implementation for each distinct type of diffusion. The simplest and most popular implementation considers a simple process where signals are transmitted from each node to every other node, by following shortest paths. This approach is often used in transportation networks to provide an estimate of the traffic handled by the nodes, assuming that the number of shortest paths that traverse a node is a good approximation for the frequency of use of the node. The centrality is then estimated by counting how many times a node is crossed by those paths. The higher the count, the more traffic is controlled by the node, which is therefore more influential in the network.

Given two nodes, there may be more than one shortest path between them in the network, all having the same length. For instance, if nodes X and Y are not connected to each other but have two common neighbors S and T, there are two distinct shortest paths of length two running from X to Y: $X-S-Y$ and $X-T-Y$. Let σ_{hj} be the total number of shortest paths from h to j and $\sigma_{hj}(i)$ the number of these shortest paths that pass through node i. The betweenness of i is defined as

$$b_i = \sum_{h \neq j \neq i} \frac{\sigma_{hj}(i)}{\sigma_{hj}}. \tag{3.3}$$

In Eq. (3.3) the sum runs over all pairs of vertices h and j, distinct from i and from each other. If no shortest path between h and j crosses i [$\sigma_{hj}(i) = 0$], the contribution of the pair (h, j) to the betweenness of i is 0. If all shortest paths between h and j cross i [$\sigma_{hj}(i) = \sigma_{hj}$], the contribution is 1. If a node is a *leaf* (i.e. it has only one neighbor), it cannot be crossed by any path. Therefore its betweenness is zero. Since the potential contributions come from all pairs of nodes, the betweenness grows with the network size.

Let us work through the example in Figure 3.1(a). For node **1**, the only pair of nodes that has a shortest path going through this node is $(\mathbf{2}, \mathbf{4})$. However, there are two shortest paths of equal length between **2** and **4**: the other path goes through node **3** and not **1**. Therefore the betweenness of node **1** is $1/2$. Next, consider node **3**. The shortest paths between the three node pairs $(\mathbf{1}, \mathbf{5})$, $(\mathbf{2}, \mathbf{5})$, and $(\mathbf{4}, \mathbf{5})$ go through **3**. As we observed earlier, there are two equivalent shortest paths between nodes **2** and **4**, only one of which goes through **3**, contributing $1/2$ to the sum. The total gives a betweenness centrality of 3.5 for node **3**. The remaining nodes **2**, **4**, and **5** have no shortest paths going through them, therefore their betweenness is zero.

A node has high betweenness if it occupies a special position in the network, such that it is an important station for the communication patterns running through the network. For that to happen, it is not necessary to have many neighbors. Generally we observe a correlation between the degree of a node and its betweenness, so that well-connected nodes have high betweenness and vice versa [Figure 3.1(a)]. However, there are many exceptions. Nodes bridging different regions of a network typically have high betweenness, even if their degree is low, as illustrated in Figure 3.1(b).

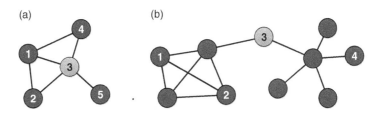

Fig. 3.1 Illustrations of node betweenness centrality. (a) The orange node has high degree ($k_3 = 4$) as well as high betweenness ($b_3 = 3.5$). (b) The orange node has low degree ($k_3 = 2$) but keeps the network connected, acting as the only bridge between nodes in the two subnetworks. For example, the shortest path between nodes **1** and **2** does not go through the orange node, but the path between **1** and **4** does. In fact, all the shortest paths between the four nodes in one subnetwork and the five nodes in the other subnetwork go through the orange node. Therefore its betweenness is $b_3 = 4 \times 5 = 20$.

The concept has a straightforward extension to links. The betweenness centrality of a link is the fraction of shortest paths among all possible node couples that pass through that link. Links with very high betweenness centrality often join cohesive regions of the network, called *communities*. Therefore betweenness can be used to locate and remove those links, allowing for the separation and consequent identification of communities (Chapter 6).

The betweenness centrality depends on the size of the network. If we wish to compare the centrality of nodes or links in different networks, the betweenness values should be normalized.

For node betweenness, the maximum number of paths that could go through node i is the number of pairs of nodes excluding i itself. This is expressed by $\binom{N-1}{2} = \frac{(N-1)(N-2)}{2}$. The normalized betweenness of node i is therefore obtained by dividing b_i in Eq. (3.3) by this factor.

NetworkX has functions to compute the normalized betweenness centrality of nodes and links:

```
nx.betweenness_centrality(G)          # dict nodes ->
                                      # betweenness centrality
nx.edge_betweenness_centrality(G)  # dict links ->
                                      # betweenness centrality
```

3.2 Centrality Distributions

Before the advent of online social media, the social networks that one could study were typically built through personal interviews and surveys, which could not involve very many people in a reasonable time frame. As a result, the networks consisted of only a few dozen nodes. On such small networks it makes sense to differentiate individual nodes and ask questions such as "what is the most important node of the network?" Nowadays we handle much larger graphs. For instance, the social network of Facebook friendships involves two billion individuals, including many prominent people like famous artists, sports celebrities, politicians, business people, and scientists, among others. However, no matter how popular, each of them can only be connected to a small portion of the entire network.

To better understand how centrality is distributed among the many nodes in large networks, we need to take a *statistical* approach. In this way we can focus on classes of nodes and links sharing similar features, rather than on single elements of the network. For example, we can group together all nodes having similar values of degree centrality. The statistical distribution of a centrality measure tells us how many elements — nodes or links — have a certain value of centrality, for all possible values. Figure 3.2 shows, for example, the distribution of node degree in a small network. In large networks, this is a useful tool to identify the classes of elements: by examining the distribution, we can see if there are notable values or groups of values and classify the elements accordingly. The range of the distribution also reveals the heterogeneity of the network elements with respect to a specific centrality measure of interest: for example, if node degrees span many orders of magnitude, from units to millions, then the network is very heterogeneous with respect to degree. Such heterogeneity has implications both on the structure of the network and on its function, as we shall see.

Box 3.1 defines probability distributions and shows how to calculate them. To examine the probability distributions of centrality measures in real networks, let us focus on two systems: a network of Twitter users and a network of math articles on Wikipedia (en.wikipedia.org/wiki/Category:Mathematics). In the Twitter network, nodes are users and a directed link from user Alice to user Bob indicates that Bob retweets (some of)

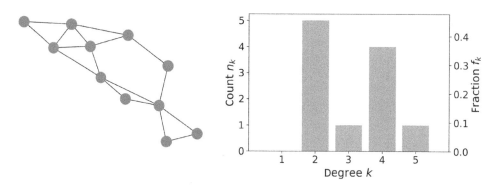

Fig. 3.2 Histogram representation of the degree distribution of a small network. First a list with the degree of each node is generated. The heights of the histogram bars are given by the counts n_k of nodes with each degree k. The relative frequency of occurrence f_k is defined as the fraction of all the nodes with degree k. The values of f_k are also shown.

Box 3.1   **Statistical Distributions**

The *histogram* or *distribution* of a quantity (e.g. a centrality measure) is a function that counts the number of observations (e.g. nodes) having different values of the quantity. If the quantity of interest is discrete (e.g. integer), for each value v we count the number n_v of observations having that value. So, the sum of n_v over all values is the total number of observations: $\sum_v n_v = N$. The result is plotted as a series of consecutive bars, one for each value, whose height is n_v.

To compare histograms of different sets of observations, it is common to divide n_v by the total number of observations N, yielding the *relative frequency* $f_v = n_v/N$. The sum of all relative frequencies is 1, regardless of the number of observations. For the node degree, the normalization is obtained by dividing by the total number of nodes (Figure 3.2). The relative frequency f_v is then the fraction of nodes with degree v.

In the limit of infinitely many observations, f_v converges to the *probability* p_v that an observation takes value v. In this limit, the histogram becomes a *probability distribution*. Any real-world network has a finite number of nodes and links, so it is impossible to reach the infinite limit, and the histogram is only an approximation of the probability distribution. However, if the network is large enough, say millions of nodes, we can treat it as a probability distribution for practical purposes.

While some centrality measures, such as the degree of a node, take integer values, other do not. For example, betweenness centrality values are not necessarily integer. In these cases, instead of counting observations for specific values, we can divide the range of values into disjoint intervals, or *bins*. Then we can similarly count the number of observations falling within each bin. This binning technique can be used whenever we are interested in ranges of values, even if the values are integer. For instance, a histogram of individual wealth may count how many individuals have a yearly income within brackets, like $0–50k, $50k–100k, $100k–200k, and so on.

The complementary cumulative distribution function, or simply *cumulative distribution* $P(x)$ of a variable, gives the probability that an observation has a value larger than x. To compute $P(x)$, we sum the relative frequencies (or probabilities) of all the values of the variable to the right of the value x: $P(x) = \sum_{v \geq x} f_v$. The cumulative distribution is often used when the range of variability is very broad, as is the case for several centrality measures in real-world heterogeneous networks. Since high values of the variables are rare, the standard distribution has a noisy tail. The cumulative distribution effectively averages out the noise.

the content originally broadcast by Alice. Wikipedia nodes are pages, and links are hyperlinks leading from one page to another. Both networks are directed. The Twitter network that we consider has 18,470 nodes and 48,365 links (average in-degree 2.6). The Wikipedia network has 15,220 nodes and 194,103 links (average in-degree 12.8). Such low values of average degree compared to system size indicate that both networks are sparse (i.e. very few pairs of nodes are connected by links). This is a common feature of many real networks (Table 1.1).

Let us focus on the distribution of degree centrality. In Figure 3.3 we show the *cumulative degree distribution* of both networks (Box 3.1). The curves span several orders of magnitude. In such cases one says that the distributions are *broad* or have a *heavy tail*,

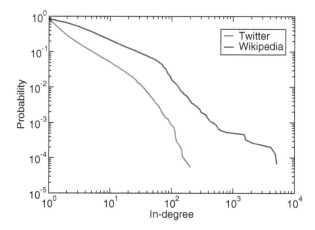

Fig. 3.3 Cumulative degree distributions of Twitter and Wikipedia networks, shown on a log–log plot. Both graphs are directed, so we show the in-degree distributions. The maximum in-degree is 204 for Twitter and 5171 for Wikipedia. The curves are plotted using the logarithmic scale because they span several orders of magnitude.

Box 3.2   **The Logarithmic Scale**

When plotting a curve that includes very small and very large values on one or both axes, differences between small values are indistinguishable. A solution is to plot in *logarithmic scale*: instead of using the original values as axis coordinates, we use their *logarithms*. This way a large range of values spanning many orders of magnitude can be represented effectively: small differences are amplified in the range of small values and large differences are compressed in the range of large values. We use the logarithmic scale to plot heavy-tailed distributions of network centrality measures. Since both the centrality values and the probability values span several orders of magnitude, the logarithmic scale is used on both x and y axes. We call such diagrams *log–log plots*.

---

where the tail is the right portion of the distribution, reaching to the largest values of the variable. It is customary to use the cumulative distribution when a measure has such a broad variability range. Also, heavy-tailed distributions are more effectively plotted in double *logarithmic scale*, or *log–log* scale (Box 3.2), as we have done in Figure 3.3, to be able to resolve the shape of the distribution at different orders of magnitude.

Heavy-tailed degree distributions display a large heterogeneity in the degree values: while many of the nodes have just a few neighbors, some others have many neighbors, which gives them a prominent role in the network. These nodes are called *hubs*. Many natural, social, information and man-made networks have heavy-tailed degree distributions with highly connected hubs. One way to measure the breadth of the degree distribution is to compute the *heterogeneity parameter*, which compares the variability of the degree across nodes to the average degree.

To formally define the heterogeneity parameter κ (Greek letter "kappa") of a network's degree distribution, we need to introduce the *average squared degree* $\langle k^2 \rangle$, which is the average of the squares of the degrees:

$$\langle k^2 \rangle = \frac{k_1^2 + k_2^2 + \cdots + k_{N-1}^2 + k_N^2}{N} = \frac{\sum_i k_i^2}{N}. \tag{3.4}$$

The heterogeneity parameter can be defined as the ratio between the average squared degree and the square of the average degree of the network [Eq. (1.5)]:

$$\kappa = \frac{\langle k^2 \rangle}{\langle k \rangle^2}. \tag{3.5}$$

For a normal or narrow distribution with a sharp peak at some value, say k_0, the distribution of the squared degrees is concentrated around k_0^2. Therefore $\langle k^2 \rangle \approx k_0^2$ and $\langle k \rangle \approx k_0$, yielding $\kappa \approx 1$. For a heavy-tailed distribution with the same average degree k_0, $\langle k^2 \rangle$ blows up because of the large degree of the hubs, so that $\kappa \gg 1$.

If the degree distribution is concentrated around a typical value, there is no heterogeneity and the parameter is typically close to one.[1] If the degree distribution is broad instead, the heterogeneity parameter is heavily inflated by the largest degrees of the hubs, and may take large values. The more hubs there are, the larger the heterogeneity. As we shall see, heterogeneity plays a key role in the structure of a network, and in the dynamics of some processes running on it.

If a network is directed, like our Wikipedia and Twitter graphs, we have to consider two distributions, the *in-degree* and *out-degree* distributions, defined as the probability that a randomly chosen vertex has a given in- or out-degree, respectively. In this case, the definition of hub may refer to either the in-degree or the out-degree. For instance, a Web page may have many other pages linking to it (large in-degree), but it may itself link to just a few pages (low out-degree), or vice versa. In several directed networks the two measures are *correlated*, so nodes with large (small) in-degree also have large (small) out-degree. We will return to the discussion of degree in directed as well as weighted networks in Chapter 4. Table 3.1 reports some basic numbers characterizing the degree distributions of various networks.[2]

One can of course analyze the distributions of other properties besides the degree. It turns out that the degree is usually correlated with other centrality measures. So, hubs typically rank among the most central nodes with respect to diverse criteria. There are exceptions as well. As we have seen in Figure 3.1, a node may have a large betweenness centrality if it connects different areas of the network, whether or not it has high degree.

In Figure 3.4 we show the cumulative betweenness distributions for our networks. Just like the degree distributions, they too span multiple orders of magnitude.

---

[1] An alternative definition in the literature compares the heterogeneity parameter with $\langle k \rangle$ rather than with 1.

[2] Datasets for these networks are available in the book's GitHub repository: github.com/CambridgeUniversity Press/FirstCourseNetworkScience

Table 3.1 Basic variables characterizing the degree distribution of various network examples: average degree, maximum degree, and heterogeneity parameter. The networks are the same as in Tables 1.1 and 2.1, their numbers of nodes and links are listed as well. For directed networks we report the maximum in-degree, and the heterogeneity parameter is computed on the in-degree distribution

Network	Nodes (N)	Links (L)	Average degree $(\langle k \rangle)$	Maximum degree (k_{max})	Heterogeneity parameter (κ)
Facebook Northwestern Univ.	10,567	488,337	92.4	2,105	1.8
IMDB movies and stars	563,443	921,160	3.3	800	5.4
IMDB co-stars	252,999	1,015,187	8.0	456	4.6
Twitter US politics	18,470	48,365	2.6	204	8.3
Enron email	87,273	321,918	3.7	1,338	17.4
Wikipedia math	15,220	194,103	12.8	5,171	38.2
Internet routers	190,914	607,610	6.4	1,071	6.0
US air transportation	546	2,781	10.2	153	5.3
World air transportation	3,179	18,617	11.7	246	5.5
Yeast protein interactions	1,870	2,277	2.4	56	2.7
C. elegans brain	297	2,345	7.9	134	2.7
Everglades ecological food web	69	916	13.3	63	2.2

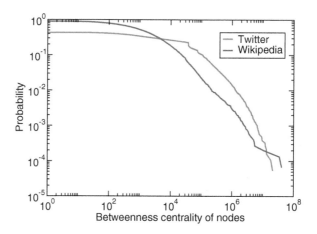

Fig. 3.4 Cumulative distribution of node betweenness centrality for Twitter and Wikipedia, shown on a log–log plot. We considered both networks as undirected. For Wikipedia we computed the betweenness only on its giant component, which includes over 98% of the nodes. The Twitter graph is connected.

Hubs, when present, are the single most important feature of a network. They are the pillars of its structure and the drivers of the processes running on it. In the next sections we present some remarkable consequences of the presence of hubs.

3.3 The Friendship Paradox

Suppose you are looking for the person who has the largest number of friends, among a group of N people for whom you only have a directory of phone numbers. If you just call one of the numbers, chosen at random, the chance that you picked the right person is $1/N$. What if you ask them about one of their friends? It may seem that you are just selecting another individual at random, like before, and that the chance that the friend is the right person is the same. But that is not the case. To see why, consider the small social network in Figure 3.5. The most connected individual is Tom, who has four friends. If you question a random individual, there is one possibility out of seven that you pick Tom. However, if you select a random friend of a random individual, the probability that you bump into Tom turns out to be $5/21 \approx 24\%$, which is quite a bit larger than $1/7 \approx 14\%$. We conclude that it is easier to find people through their friends than by random search. But why?

Roughly speaking, if someone has many friends, she has a far greater chance of being mentioned by someone than if she had just a few. Reaching out to someone's friends actually means choosing links instead of nodes. When we go for the nodes, each of them has the same probability of being selected, regardless of their degree. When we go for the links, the larger the number of neighbors of a node, the higher the probability it will be reached. In our network of Figure 3.5 there are four possible channels leading to Tom, so it is far easier to reach him than Mary or Tara, who only have one friend.

The chances of hitting a hub increase if you move from the circle of neighbors to that of the neighbors of neighbors, and so on. This is because the number of links to follow increases at each step, so it becomes more likely that one of them is attached to a hub. This property can be used to our advantage. There are many situations in which identifying the hubs of the network could be helpful. For instance, during an epidemic outbreak, the individuals with the largest number of contacts are potential big spreaders and it would be important to isolate and/or vaccinate them to contain the disease. In such a scenario, one

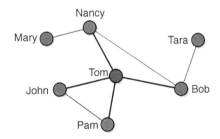

Fig. 3.5 Friendship Paradox. By selecting a random link instead of a random node, Tom can be "found" much more easily than Mary, because he has four friends (John, Pam, Bob, and Nancy), whereas she has only one (Nancy). It is far more likely to bump into a hub than into a node with low degree when following connections at random. This is the underlying reason why our friends have more friends than us, on average.

could select people randomly and get in touch with some of their friends, as they have a higher probability of being hubs than the pool of selected individuals. We will revisit this topic in Chapter 7.

The difference in the selection of links versus nodes has another peculiar implication. Let us choose an actor in our network, say Nancy. She has three friends: Bob, Mary, and Tom. They have in total $3 + 1 + 4 = 8$ friends, which gives an average of $8/3$. If we repeat this calculation for all other nodes, we find that the average number of neighbors of the neighbors of a node is $17/6 = 2.83$. However, the average degree of the network is $(1 + 3 + 3 + 1 + 4 + 2 + 2)/7 = 16/7 = 2.29$. This is typical: the average degree of the neighbors of a node is larger than the average degree of the node. In other words, our friends have more friends than us, on average. This is known as the *Friendship Paradox*.

Our example helps us uncover the origin of the paradox. When we compute the average degree of a node, each node's degree appears only once in the sum. In contrast, when we compute the average degree of the neighbors of a node, and we repeat the procedure for all nodes, each node will appear as many times as its degree in the partial sums. In our example, Tom's degree will be counted four times because he is in the list of friends of four people. This boosts the value of the neighbors' average degree, which ends up being larger than the average degree. The Friendship Paradox is thus due to *sampling*. The two averages are computed by sampling the node degrees differently: uniformly for average degree, proportionally to the degree for the neighbors' average degree.

The broader the degree distribution, the stronger the effect of the Friendship Paradox. When all nodes have approximately the same degree, the two values are similar to each other. In networks with heavy-tailed distributions, like the ones featured in Figure 3.3 (and like typical social networks), the effect is very pronounced because of the super-connected hubs.

3.4 Ultra-Small Worlds

The hubs of a network are not only easy to find, as we have just seen; they are also in great demand. If we want to transmit a signal from one node of the network to another along the shortest route, the signal is likely to pass through one or more hubs. You have probably experienced this in your air travel: when you want to fly from airport **A** to airport **B**, and there is no direct flight between **A** and **B**, you are forced to make at least one connection at some hub airport **C**. In many cases, one connection is enough, so the trip from **A** to **B** requires only two flights: $\mathbf{A} \rightarrow \mathbf{C}$ and $\mathbf{C} \rightarrow \mathbf{B}$.

In Chapter 2 we have seen that many real networks are *small worlds* (i.e. one can go from every node to any other node with a small number of steps). In a network with hubs, we expect that the average distance between any two nodes is shorter compared to a network with the same number of nodes and links, but no hubs. In fact, networks with broad degree distributions often have the so-called *ultra-small world* property, indicating that

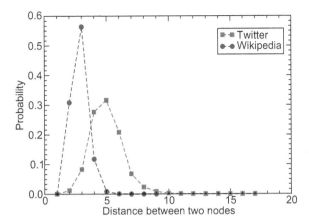

Ultra-small worlds. The distributions of distance between nodes are peaked at very low values for both Twitter and Wikipedia. This is due to the presence of hubs, which shrink the distance between most pairs of nodes, as shortest paths run through them. Distances are computed by ignoring the direction of the links.

the distances between nodes are very short. In Figure 3.6 we plot the distribution of the distances between any two nodes for the reference networks we have been using: Twitter and Wikipedia. Both distributions are strongly peaked, so there is very little variability among the distances. The peak values are extremely small compared to the system sizes (five for Twitter, three for Wikipedia), indicating that both networks are ultra-small worlds. This is a trademark feature of many real networks.

3.5 Robustness

A system is *robust* if the failure of some of its components does not affect its function. For instance, an airplane keeps flying if one of its engines stops working. In general, robustness depends on which components fail and on the extent of the damage.

How to define the robustness of a network? Nodes can describe a broad variety of entities, such as people, routers, proteins, neurons, websites, and airports. In such a high-level representation, it is not straightforward to define the failure of a node, which depends on the specific type of network. But if we assume that a node stops working somehow, we can ask how the structure and consequently the function of the network changes without that node and all of its links.

In Chapter 2 we defined what it means for a network to be connected — all nodes are reachable from each other. We also saw that if a network is not connected, it has two or more connected components. Connectedness is an important network property that typically affects its function. If the Internet were not a connected graph, it would be impossible to send signals (e.g. emails) between routers belonging to different components. Therefore,

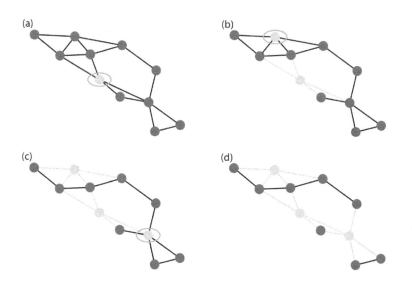

Fig. 3.7 Network robustness. Effect of a sequence of deletions of nodes and their incident links. In each diagram the deleted node is highlighted by a circle. Deleted nodes and their incident links are colored in gray. After three nodes are deleted (d), the network breaks into three components disconnected from each other.

one way to define and measure the robustness of a network is to observe how the removal of a node and its links affects the connectedness of the system (Figure 3.7). If the system remains connected, we can assume that it will keep working fine, to some extent. However, a breakup of the network into disconnected pieces would signal severe damage that might compromise its function.

The standard robustness test for networks consists of checking how the connectedness is affected as more and more nodes are removed, along with all of their adjacent links. To estimate the amount of disruption following node removal, scholars compute the relative size of the giant component (i.e. the ratio of the number of nodes in the giant component to the number of nodes initially present in the network). Let us suppose that the initial network is connected. In this case the giant component coincides with the whole network, so its relative size is one. If the removal of a subset of nodes does not break it into disconnected pieces, the proportion of nodes in the giant component just decreases by the fraction of removed nodes. If, however, the node removal breaks the network into two or more connected components, the size of the giant component may drop substantially. As the fraction of removed nodes approaches one, the few remaining nodes are likely distributed among tiny components, so the proportion of nodes in the giant component is close to zero.

Figure 3.8 illustrates the outcomes of robustness tests on the OpenFlights World network. When nodes are removed randomly, the process simulates the *random failure* of network elements. We observe that the relative size of the giant component decreases very slowly. This is due to the presence of hub nodes, which keep the structure connected. As

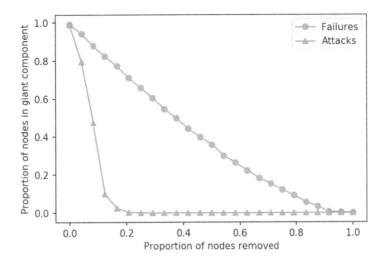

Network robustness. Fraction of nodes in the giant component as a function of the fraction of nodes removed from the OpenFlights World network. We show what happens if nodes are removed at random (random failures), or prioritized based on degree (targeted attacks).

long as a sufficient number of hubs survive, the system remains largely connected. Since we are removing nodes at random, the probability of hub failure is low, because they are statistically rare compared to other nodes. The plot also shows what happens when nodes are removed in decreasing order of their degree (i.e. the hubs are targeted first). In this case, the system suffers a major disruption almost immediately, and is totally fragmented when about 20% of the nodes are eliminated. Targeting high-degree nodes is an example of *attack*, as one aims to maximize damage by removing central nodes. We conclude that many real networks, which have central hubs, are pretty robust to random failures, but quite vulnerable to attacks.

3.6 Core Decomposition

We briefly mentioned the *core–periphery structure* of many networks in Section 2.1. When analyzing or visualizing a large network, it is often useful to focus on its denser portion (core).

The degree of each node can be used to separate a network into distinct portions, called *shells*, based on their position in the core–periphery structure of the network. Low-degree outer shells correspond to the periphery. As they are removed, or peeled away, what remains is a denser and denser inner subnetwork, the *core*. We start with singletons (zero-degree nodes), if there are any. Then we remove all nodes with degree one. Once there are none left, we start removing nodes with degree two, and so on. The last group of nodes to be removed is the innermost core.

Formally, the *k-core decomposition* algorithm starts by setting $k = 0$. Then it proceeds iteratively. Each iteration corresponds to a value of k and consists of a few simple steps:

1. Recursively remove all nodes of degree k, until there are no more left.
2. The removed nodes make up the *k-shell* and the remaining nodes make up the $(k + 1)$-*core*, because they all have degree $k + 1$ or larger.
3. If there are no nodes left in the core, terminate; else, increment k for the next iteration.

Core decomposition is particularly helpful in practice for filtering out peripheral nodes when visualizing large networks. In fact, most of the figures in Chapter 0 do not depict entire networks, but only portions obtained by excluding some of the periphery. For example, the $k = 1$ and $k = 2$ shells have been removed in the political retweet network shown in Figure 0.3. This filtering process is illustrated in Figure 3.9.

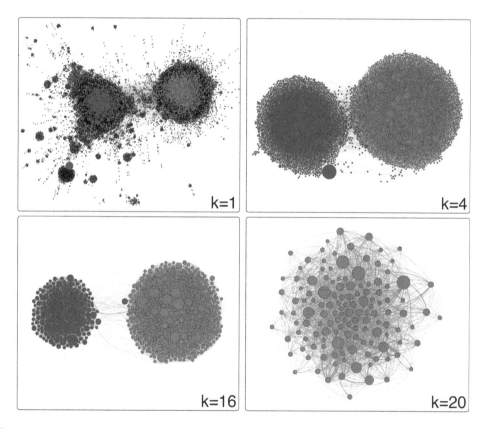

Fig. 3.9 Filtering by k-core decomposition. We start from the full Twitter political retweet network ($k = 1$). As we increase k, peripheral nodes are removed and the remaining core becomes smaller and denser. The innermost core contains only red nodes, corresponding to conservative accounts; each has at least $k = 20$ neighbors.

NetworkX has functions for core decomposition:

```
nx.core_number(G)  # return dict with core number of each node
nx.k_shell(G,k)    # subnetwork induced by nodes in k-shell
nx.k_core(G,k)     # subnetwork induced by nodes in k-core
nx.k_core(G)       # innermost (max-degree) core subnetwork
```

3.7 Summary

In this chapter we have learned about different centrality measures of network nodes and edges, and focused on the degree of nodes as an important measure that identifies hubs. A few concepts to remember:

1. The degree of a node is defined as the number of links in the graph incident on the node.
2. The betweenness of a node expresses how often it is traversed by signals propagating on the networks following shortest paths.
3. In large networks it is necessary to use statistical tools to analyze the global features of the network. The histogram provides a visual illustration of the distribution of a given attribute of nodes or links (e.g. the degree). The normalized histogram is an estimate of the probability distribution of the measure of interest.
4. The distributions of centrality measures are heterogeneous for many real networks (i.e. they span multiple orders of magnitude). In particular, the degree distribution often has a heavy tail. Nodes with large degree are called hubs.
5. The Friendship Paradox says that in a social network, your friends have more friends than you do, on average. This is due to the high probability of selecting hubs among a node's neighbors.
6. Hubs play a critical role in the structure and dynamics of a network. For instance, they shrink the distances between nodes and make the network robust against random failures, but vulnerable to targeted attacks.
7. We can decompose the network to reveal its core–periphery structure. This is accomplished by iteratively filtering out shells of low-degree nodes and focusing on the remaining, denser and denser cores.

3.8 Further Reading

Closeness centrality was introduced by Bavelas (1950). Freeman (1977) introduced node betweenness and Brandes (2001) developed the algorithm commonly adopted to calculate it. Link betweenness, introduced in an unpublished technical report by Anthonisse, is well described by Girvan and Newman (2002), who applied the measure to detect and remove links connecting network communities to each other, so that the latter can be separated and identified (Section 6.3.1). Statistical distributions are nicely presented in the book by Freedman *et al.* (2007).

An accessible introduction to networks and their hub structure is offered by Barabási (2003). Albert *et al.* (1999) discovered the first large network with a heavy-tailed degree distribution, namely the Web graph. Many other real-world networks were subsequently found to have the same property (Barabási, 2016).

The Friendship Paradox was exposed by Feld (1991). Ultra-small worlds were discovered by Cohen and coworkers (Cohen and Havlin, 2003; Cohen *et al.*, 2002, 2003). The first study of network robustness appeared in a paper by Albert *et al.* (2000). Cohen *et al.* (2000, 2001) have authored classic theoretical studies on robustness.

The application of k-core decomposition to network visualization is due to Batagelj *et al.* (1999), Baur *et al.* (2004), and Beiró *et al.* (2008).

Exercises

3.1 Go through the Chapter 3 Tutorial on the book's GitHub repository.[3]

3.2 Assume you have a graph with 100 nodes and 200 links. What is the average degree of nodes in this network?

3.3 Consider a network formed by 250 students in a dormitory. The links in this network represent room-mate relationships: two nodes are connected if they are currently room-mates. In this dorm, the rooms are mostly double occupancy with a few triples and quads.
1. Is this graph connected?
2. What is the mode (most frequent value) of the node degree distribution?
3. How many nodes are in the largest clique?
4. Would you expect this graph to have any hubs?

3.4 In NetworkX, how can you find a node with the largest degree centrality in a network? And how would you also get the degree of that node?

3.5 Assume you have a NetworkX graph G of employees. The node names are employee IDs, and the nodes have attributes for full name, department, position, and salary. Which of the following will give you the salary for the employee with ID 5567?
   a. `G.node(5567)('salary')`
   b. `G[5567]['salary']`
   c. `G.node[5567]['salary']`
   d. `G(5567)('salary')`

3.6 You have a NetworkX graph G and you are about to draw it with the following command: `nx.draw(G, node_size=node_size_list)`. Which of the following is a correct way to obtain `node_size_list` so that the nodes are sized according to their degree?

a. `node_size_list = [G[n] for n in G.nodes]`

b. `node_size_list = G.degree()`

c. `node_size_list = [G.degree() for n in G.nodes]`

d. `node_size_list = [G.degree(n) for n in G.nodes]`

e. `node_size_list = [d for d in G.degree()]`

3.7 An academic collaboration network is one type of social network. In such a network, a node with degree two means that:

   a. A scholar has coauthored a paper with one other scholar

   b. A scholar has coauthored publications with two other scholars

   c. A scholar has authored two publications

   d. A publication was coauthored by two scholars

3.8 In a social network, which of the following would one expect to be true about the degrees of its nodes?

   a. Most nodes connect to a single, large hub

   b. A variety of degrees is to be found

   c. All nodes have more or less the same degree

   d. All nodes have very high degree

3.9 What property does a network need to have in order for closeness centrality to be well-defined?

3.10 Provide examples of networks such that:

   1. The node with the highest degree is not the one with largest closeness

   2. The node with the highest betweenness is not the one with largest closeness

3.11 Consider the network in Figure 3.10 in order to answer the next few questions. For each question, in case of a tie, answer with all the tied top nodes.

   1. Which node has the highest degree centrality?

   2. Which node has the highest betweenness centrality?

   3. Which node has the highest closeness centrality?

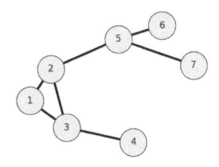

Fig. 3.10   An undirected, unweighted network.

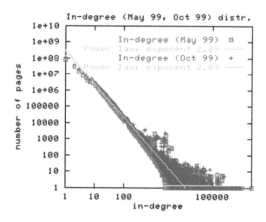

Fig. 3.11   Histogram of Web in-degree on a log–log scale. Reprinted from Broder *et al.* (2000) with permission from Elsevier.

3.12 Suppose we want to build a connected network with 10 nodes and average degree 1.8, such that the heterogeneity parameter is largest. What does the graph look like?

3.13 For each of the following variables, state whether or not you would expect to see a heavy-tailed distribution, and why:
1. The shoe size of UK adults
2. US household income
3. Node degree in the Twitter social network
4. Pairwise distance in the Wikipedia network

3.14 If people's heights followed a heavy-tailed distribution, would you be surprised to see a 30-ft (9-m) tall person in the street?

3.15 The plot in Figure 3.11 comes from a study of 200 million Web pages and 1.5 billion links between them (Broder *et al.*, 2000). It is a log–log plot of the number of pages (*y*-axis) with a given number of in-links (*x*-axis):
1. Approximately how many pages have only one other page linking to them?
2. Approximately how many pages have 10 other pages linking to them?
3. Approximately how many pages have 100 other pages linking to them?

3.16 Consider a social network where a connection represents a sexual relationship. Read the report by Liljeros *et al.* (2001) about a study of such a network based on a sample of 4781 Swedes. (If you do not have access to the journal through your institution, you can download a preprint of the paper at https://arxiv.org/abs/cond-mat/0106507.) What is the maximum degree in this network? What does it mean? If you consider the subnetworks with nodes corresponding to males and females, respectively, do they have the same degree distribution? Why or why not?

3.17 A common use of the word "hub" in everyday speech is to describe airports that serve many routes (direct flights). Load the OpenFlights US flight network into a NetworkX graph to answer the following questions:

1. What is the average number of routes served by each airport in this network?
2. What are the top five airports in terms of number of routes?
3. How many airports in this network serve only a single route?
4. Which airport has the highest closeness centrality?
5. Which airport has the highest betweenness centrality?
6. Compute the heterogeneity parameter of this network.

3.18 Load the Wikipedia mathematics network into a NetworkX digraph in order to answer the following questions:
1. Compute the average in-degree and average out-degree of this network. What do you notice? Why?
2. Which node has the highest in-degree?
3. Which node has the highest out-degree?
4. In this graph, which is greater: the maximum in-degree or the maximum out-degree? Would you expect this to be the same for other Web graphs? Why?
5. Compute the heterogeneity parameter for this graph's in-degree distribution.
6. Compute the heterogeneity parameter for this graph's out-degree distribution.

3.19 Write a Python function that accepts a NetworkX graph and a node name and returns the average degree of that node's neighbors. Use this function to compute this quantity for every node in the OpenFlights US network and take the average. Does the Friendship Paradox hold here (i.e. is the average degree of nearest neighbors greater than the average node degree)?

3.20 Are there networks such that the average number of neighbors of a node's neighbors matches the average degree? If there are, what property must they have?

3.21 Are networks with heavy-tailed degree distributions more vulnerable to random or targeted attacks? And what about grid-like networks of similar size?

3.22 If one seeks to disrupt a network by removing nodes and/or edges in an effort to disconnect it and/or increase the average path length, an obvious strategy is to attack the hubs. Which of the following is another deleterious criterion for selecting targets? Explain your answer.
a. Nodes with high clustering coefficient
b. Nodes with low degree
c. Nodes with high closeness centrality
d. Nodes/edges with high betweenness centrality

3.23 Consider two nodes of equal degree on some network: one with high clustering coefficient and one with low clustering coefficient. All else being equal, which of the two would you intuit to be a better target if you were seeking to disrupt the network?

3.24 The `socfb-Northwestern25` network in the book's GitHub repository is a snapshot of Northwestern University's Facebook network. The nodes are anonymous users and the links are friend relationships. Load this network into a NetworkX graph in order to answer the following questions. Be sure to use the proper graph class for an undirected, unweighted network.

1. What proportion of nodes have degree 100 or greater?
2. What is the maximum degree for nodes in this network?
3. Users in this network are anonymized by giving the nodes numerical names. Which node has the highest degree?
4. What is the 95th percentile for degree (i.e. the value such that 95% of nodes have this degree or less)?
5. What is the mean degree for nodes in this network? Round to the nearest integer.
6. Which of the following shapes best describes the degree distribution in this network? You can obtain the answer visually using histograms, or just with statistics.

   a. Uniform: node degrees are evenly distributed between the minimum and maximum

   b. Normal: most node degrees are near the mean, dropping off rapidly in both directions

   c. Right-tailed: most node degrees are relatively small compared to the range of degrees

   d. Left-tailed: most node degrees are relatively large compared to the range of degrees

4 Directions and Weights

link: (*n.*) a relationship between two things or situations, especially where one thing affects the other.

Many real-world networks are directed and weighted. We saw for instance in Chapter 0 that food webs connect species with directed and weighted links that represent the direction and amount of prey consumed by the predator. Other familiar examples include Wikipedia and the Web at large, where hyperlinks are weighted by click traffic; every app where we rate products and services, from books to movies and from drivers to songs; social networks stemming from Twitter, where friend/follower links are weighted by retweets, quotes, replies, and mentions; and even Facebook, where friend interactions are weighted by comments, "likes," and reshares. Many of these networks are built on top of Internet and Web technologies. In this chapter you will become familiar with some of these networks and protocols.

4.1 Directed Networks

In the networks that we have discussed until this point, the direction of the links is not important. In social networks, we often assume — even if it is not always true — that friendship is symmetric: if Alice is a friend of Bob's, then Bob is equally a friend of Alice's. In the Internet, packets travel in both directions between two routers, or two autonomous systems. Cars travel both ways on most roads, and for every flight from New York to Rome there is a flight from Rome to New York. In many networks these symmetries do not hold; a link has a particular direction, and may not be reciprocated. For instance, if Charlie follows Donna on Twitter, Donna may not follow Charlie back. As mentioned in Chapter 1, a *directed link* has a *source* node and a *target* node. The directed link is typically represented as an arrow pointing from source to target. We call a network with directed links a *directed network*. While in an undirected network each node has a degree, in a directed network each node has an *in-degree* (number of incoming links) and an *out-degree* (number of outgoing links).

We most often observe directed links in communication and information networks. Familiar examples include email (see Figure 0.4) and Wikipedia (see Figure 0.5). Another notable example comes from science. Scientists always build upon what was done before. When they publish their findings, they refer to related work in previous publications. The resulting *citation network* is a prototypical example of an information network. Nodes

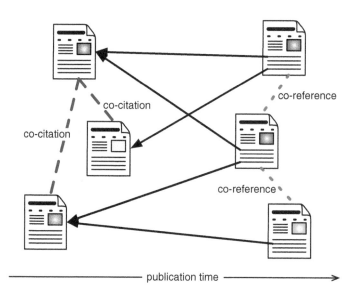

Fig. 4.1 A citation network. The citation links are shown as solid black arrows. Undirected *co-citation* and *co-reference* links induced by the citation network are also shown as dashed blue and red lines, respectively (Section 4.6). Note that in a citation network, links should always point backward in time: we cannot cite a paper that has not yet been published.

represent papers. A link from a paper to another, called a *citation*, indicates some kind of relationship between the content of the papers: a shared methodology, an alternative approach to solve a problem, a previous finding that was confirmed, improved, or possibly even contradicted. Figure 4.1 provides an illustration of citation networks.

4.2 The Web

We are all familiar with the Web, where a hyperlink leads from page **A** to page **B**, while page **B** may not have a link to page **A**. Interestingly, the idea of bi-directional links had existed for decades before the *World-Wide Web* was invented. Bi-directional links were hard to realize technically on the Web, because some type of central authority would have been needed to negotiate and store the link information.

4.2.1 A Brief History of the Web

In the early 1990s, Tim Berners-Lee introduced a directed hyperlink model that was easy to realize because anyone could link from a page to another without worrying about reciprocity from the target page or even persistence; if the target page did not exist, the user would just experience a broken link. Lots of people started writing Web pages using Berners-Lee's hypertext language for Web content, and hosting websites using his communication protocol for Web browsers and servers. The Web was born.

The link was key to the success of the Web. Each page would have a Web address called a *URL* (Uniform Resource Locator) to make it easy to link from one page to another. For about 10 years, the Web grew as many organizations created websites to present information and sell products and services. But the Web was still mainly a network in which the producers of information were distinct from its consumers. Creating a website required skills that most people did not have. The introduction of online journals called "Web logs," or *blogs*, changed that. Blogs made it easier for people to create simple sites according to templates, and produce content in the form of journal (blog) entries, hosted by third-party providers. Each blog entry would have a URL to make it easy to link from one blog entry to another. Bloggers would link to each other's blogs. Blogs quickly became the fastest-growing segment of the Web, even competing with traditional media outlets. Most importantly, many people who had been mainly consumers of information became producers as well. This was one important aspect of the so-called *Web 2.0* revolution.

Just as it became easier for people to share information through blogs, it also became easy to share photos, movies, and all sorts of other media through sites like Flickr and YouTube. People could also share links, by publishing their bookmarks on tagging sites and later on social networking sites. The link became ubiquitous and familiar. The natural next step was to link people to each other, and this happened with online social networks like Friendster, Orkut, MySpace, LinkedIn, and Facebook. To lower even more the cost of creating nodes and links, the concept of a *microblog* emerged, letting people post very short messages to be broadcast to their friends. This mix of social networks and blogs, introduced by Twitter and soon copied by Facebook, has been so popular that a significant portion of the Earth's population is now part of the Web.

From a network perspective, the concept of a node has extended to represent anything with a URL, from pages to people, from sites to thoughts, from photos to songs, from movies to articles, and so on. Likewise, the concept of a link has extended because any object can point to any other: tweets link to blog entries, Wikipedia articles cite other articles and external pages, people connect with friends and favorite things, maps link to photos (and vice versa), and so on. The Web has thus grown to encompass almost every aspect of our lives.

4.2.2 How the Web Works

Let us go back to the basics of the Web. To better understand how this ubiquitous network works, and how we can gather data about it, we need to get an idea of its language and protocol. Pages are written in some version of *HTML* (HyperText Markup Language) with interactivity provided by scripting languages, such as Javascript, that can be interpreted by the browser. Details about these languages are outside the scope of this book, except for the important concept of hyperlinks between pages, that we have already discussed. HTML provides an easy way to encode a link to another page with a special *anchor tag* (`<a>`). For example, the simple code `<a href="http://npr.org/">news</a>` creates a link for the *anchor text* "news," which, when clicked by the user, will cause the browser to fetch the page at `npr.org`.

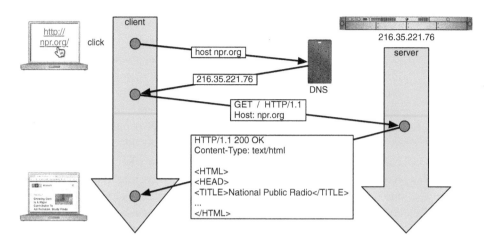

Fig. 4.2 An example of how client (browser) and server communicate through the HTTP protocol to follow a link and visit a Web page. The vertical arrows represent the direction of time.

How the fetching or downloading of pages between a *client* (the browser) and a Web server works is specified by *HTTP* (the HyperText Transfer Protocol). Similar to other Internet client–server protocols, HTTP is actually pretty simple: it prescribes how the client requests a page, and how the server responds. Figure 4.2 illustrates the protocol through an example. To first establish a connection with the server, the client must know the IP (Internet Protocol) address of the server. The URL typically specifies a server *hostname* and a *file path*. For example, in the URL http://npr.org/ the server hostname is npr.org and the file path is /. Since in this case the path is a directory, the server will look for a default file name such as index.html in that directory. To get the IP address, the browser uses a service called DNS (Domain Name Service). This is a different protocol that translates a hostname (say npr.org) into the corresponding IP address (say 216.35.221.76).

IP address in hand, the browser connects to the server. Once the connection is established, the http:// part of the URL means that the browser talks to the server using HTTP. To do this, the browser sends an HTTP request to the server and waits for an HTTP response. Request and response messages have a format consisting of a *header*, followed by an empty line, followed by an optional *body*. The request header can consist of just a couple of lines. The most common type of request is GET, which simply asks for a page (the path). In this case the request has no body, so as soon as the server receives the empty line, it responds. In other cases, such as a POST request, the body contains additional content parameters. This can be used to send input to a form. In addition to the request type and path, the header must specify the hostname. This is because a single server may often host many websites (*virtual hosting*). The response header can have several lines of information, such as the type of server, date, number of bytes returned, etc. The most important is the *response code*. For example, code 200 means "success" while code 404 means "not found." The body of the response is the actual content of the resource requested, typically the HTML code of the page.

4.2.3 Web Crawlers

Any program that uses the HTTP protocol to request content from Web servers is a Web client. The browser is the familiar tool we use to navigate the Web, allowing us to move virtually from one node to the next in this huge network of sites and pages. *Web crawlers* are programs that automatically download Web pages. Since information in the Web is scattered across billions of pages served by millions of servers around the globe, crawlers are designed to collect information that can be analyzed and mined in a central location. The primary application of crawlers is in search engines. The Web is a dynamic entity evolving at a rapid rate, therefore search engines use crawlers to stay fresh and provide current information as pages and links are added, deleted, moved, and updated. A search engine takes the information collected by a crawler and creates a data structure (an *index*) that maps content (keywords and phrases) to the pages that contain it. This way, when a user submits a query, the search engine can quickly retrieve the pages that contain those keywords. Another task of the search engine is to determine how to rank results, so that users can find quality results among millions of hits. One of the key methods to do this exploits the network structure of the Web and is presented in Section 4.3.

Other uses of Web crawlers include business intelligence, whereby organizations monitor competitors and potential collaborators; digital libraries and bibliometric systems to make scholarly work more accessible and assess its impact; webometrics tools, to evaluate the influence of institutions from their online presence; and even malicious applications, for example the harvesting of email addresses by spammers, or the collection of personal information for phishing and identity theft. Crawlers are also employed for research purposes, such as to reconstruct the structure of the Web link graph. Because crawlers are so valuable to the study of information networks, let us get an idea of how they work.

Crawlers are very complex software systems; Google founders Sergey Brin and Lawrence Page identified the Web crawler as the most sophisticated yet fragile component of a search engine. But the basic concept of a crawler is not difficult. In its simplest form, a crawler is just a breadth-first search algorithm (Section 2.5) running on the Web link graph. It starts from a set of *seed* pages (URLs) and then recursively extracts the links within them to fetch more pages, and so on. This simple description conceals technical challenges such as network connection bottlenecks, page revisit scheduling, spider traps (when meaningless URLs are generated automatically by servers), canonical URLs (to decide whether two links point to the same page), robust parsing (interpreting the often-incorrect HTML syntax of pages), and the ethics of dealing with remote Web servers.

Figure 4.3 shows the logic flow of a basic crawler. The crawler maintains a queue of unvisited URLs called the *frontier*. The list is initialized with seed URLs, which are typically a set of high-quality pages, for example from a previous crawl. In each iteration of its main loop, the crawler picks the next URL from the frontier, fetches the page corresponding to the URL through HTTP, parses the retrieved page to extract its URLs, adds newly discovered URLs to the frontier, and finally stores the page (and other extracted information, including index terms and network structure) in a repository. The crawler would stop when the frontier becomes empty, but this rarely happens in practice due to the high average out-degree of pages (on the order of 10 or more links per page across the Web). The

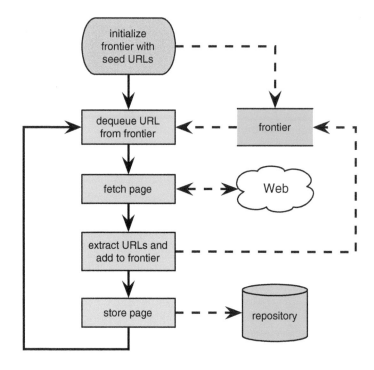

Flow chart of a basic crawler. The main data operations are shown with dashed arrows.

crawling process may be terminated when a certain number of pages have been crawled, or — as in the case of search engines — go on forever.

A typical crawler's frontier quickly becomes huge, containing many millions of unvisited links. Crawlers often employ heuristics in an attempt to prioritize links that are likely to lead to quality content. Because the frontier is commonly implemented as a first-in-first-out queue, before we visit any page at distance n from a seed, we visit all pages at distance $n - 1$ or less. This is a good strategy because, as we shall see below, the farther away we go from a quality page, the lower the chances of finding quality pages. Breadth-first crawlers seeded with large numbers of good pages can thus quickly discover many new good pages or revisit known pages to see if they have been updated since the last visit. Efficient crawlers employed by search engines use many tricks to optimize the crawling process, and devote to this task clusters with thousands of machines working in parallel around the clock. This way they crawl and index millions of pages per day, and maintain fresh search results.

4.2.4 Web Structure

The network made up of Web pages and hyperlinks is called the *Web graph*. Large-scale crawls have revealed several interesting facts about its structure. There are various (weakly) connected components. Their sizes tend to have a skewed distribution, with the largest one dominating (with more than 90% of all pages) and a great many small ones. Within the giant component, we find the largest strongly connected component.

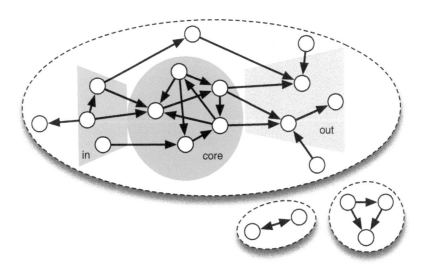

Fig. 4.4   The bow-tie structure of the Web graph. Components are highlighted by dashed ovals. The giant component has a giant strongly connected component (sometimes referred to as the "core"), an in-component ("in"), and an out-component ("out").

Recall from Section 2.3 that the *in-component* and *out-component* of the giant strongly connected component are the sets of pages from which the component can be reached and that can be reached from the component, respectively, by following directed paths. This is sometimes called a *bow-tie* structure when referring to the Web (see Figure 4.4). The relative sizes of the giant strongly connected component, in-component, and out-component vary depending on the strategies used by Web crawlers to collect network data.

Several research teams have studied the degree distribution of the Web graph based on large crawls. The average in-degree (number of links to a page) is around 10–30 links, but the standard deviation is at least an order of magnitude larger, so that the heterogeneity parameter κ is large (cf. Section 3.2). Therefore the average is not a very meaningful measure. In fact, the distribution of in-degree has a heavy tail spanning several orders of magnitude, as shown in Figure 4.5. This reveals the presence of huge hub pages in the Web, which capture a disproportionate amount of links and traffic. The skewed in-degree distribution is a robust feature of the Web, and it has not changed since the Web was only a few years old and a few hundred million pages in size.

The distribution of out-degree is more difficult to analyze. Although crawlers find pages with thousands of outgoing links, the distribution does not span as many orders of magnitude as that of the in-degree. More importantly, whereas a page with many incoming links is usually a signature of popularity, one containing too many links to other pages is typically a signature of spam behavior: so-called *link farms* created to boost a site's search ranking. Consider too that for efficiency reasons, crawlers often truncate the download of very long pages, so that measures of out-degree are unreliable.

Researchers have also used crawl data to study the average path length of the Web graph. The average path length grows very slowly with the number of nodes: as the network size

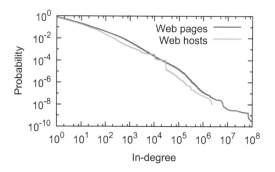

Fig. 4.5 Cumulative in-degree distribution of the Web graph. This network, based on a large crawl from August 2012, has $N = 3.6$ billion pages and $L = 129$ billion links. The distribution of in-degree is also shown for the *host graph*, in which nodes represent entire websites rather than individual pages, and a link indicates that there is at least one hyperlink between pages in two websites. In both cases we observe hub nodes with huge numbers of incoming links. (Network from webdatacommons.org based on crawl data from commoncrawl.org.)

grows by several orders of magnitude, shortest paths get only a few steps longer on average. For example, the largest strongly connected component of the network obtained from the 2012 crawl data used for Figure 4.5 has $N = 1.8$ billion pages and an average path length of less than 13 links. This *ultra-small world* structure is due to the presence of hub pages, as we discussed in Chapter 3.

These same features of the Web — heavy-tailed distributions of component size and in-degree, ultra-small world signature — have also been observed in other information networks, notably blogs and Wikipedia.

4.2.5 Topical Locality

In Section 2.1 we defined *homophily* as the tendency of similar nodes to be connected. The nodes of information networks, such as Web pages, Wikipedia articles, and research papers, are rich with content — text attributes that we can use to define and measure the *similarity* between two pages or documents. Based on the content, we can determine the topic of a page, or what it is about: two pages about sports are more similar to each other than a page about sports and one about music. Therefore we can express homophily in an information network as the capability to guess what a page is about by looking at the content of its neighbor pages. Two pages about related topics may link to each other or may have a short path connecting them. When this happens we say that the network has *topical locality*. The reason for topical locality is intuitive: when new pages, articles, or blog posts are created, the authors want to help their readers by linking to topically relevant information. As a result, links encode semantic information about the nodes.

To quantify topical locality, we can measure how likely it is that a target page within a given distance from a source page is about the same topic as the source page (Figure 4.6). We can then compare this with our expectation of running into a page about the same topic by chance, which depends on how general the topic is. Pages within one or two links from

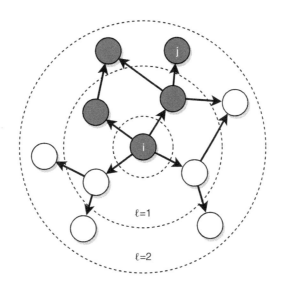

Fig. 4.6 Illustration of topical locality. Half of the pages at distance $\ell = 1$ from page i and one-third of the pages at distance $\ell = 2$ are about the same topic (shown in blue).

a source page are orders of magnitude more likely to be about the same topic as the source page, compared to random pages.

In practice, one way to measure topical locality is to use *text similarity* as a proxy for topical relatedness. Text similarity is based on the co-occurrence of keywords in two pages or documents. The more keywords are shared by two pages, the stronger the evidence that the two pages are about the same topic. A commonly used text similarity measure is the *cosine similarity*, described in Box 4.1. To quantify topical locality, one can perform a breadth-first crawl up to some distance from one or more seed pages about some topic and measure the similarity between a crawled page and a seed page, averaging across all seed pages and all pages in the crawl. Using this methodology, we can plot the similarity as a function of the distance between two pages, and observe that neighbor pages tend to be more similar than distant pages (Figure 4.7). Naturally, as we browse away from a page, we are less likely to encounter related pages; this phenomenon is called *topic drift*.

Topical locality is a kind of relationship between the structure of information networks and the content of the nodes — what the network tells us about the content, and vice versa. If we start from "good" seed pages and do not stray too far, we are likely to find other quality pages. This is one reason why search engine crawlers employ breadth-first search algorithms. Topical locality is also why it makes sense to "browse" the Web — if topical locality did not exist, would you ever click on a Web link?

There are other important hints about the content of pages that one can glean from the network structure. For example, the local network neighborhood of a page may be a signature of spam content, as when the only links to a page originate from nodes that have no incoming links themselves. Conversely, when a page has many incoming links from pages with high in-degree, that may be a clue about the quality or prestige of the page content; more on this in the next section.

Cosine Similarity

In information retrieval and text mining, one often needs to measure the similarity between two documents, Web pages, blurbs of text, or tag clouds. Let us represent each document d as a high-dimensionality vector, with a dimension associated with each term in the vocabulary: $\vec{d} = \{w_{d,1}, \ldots, w_{d,n_t}\}$, where $w_{d,t}$ is the *weight* of term t in d, and n_t is the total number of terms. Deep learning techniques based on artificial neural networks lead to similar vector representations, except that each dimension corresponds to an abstract concept rather than a word or tag. There are various ways to calculate the weights. Typically they are proportional to the frequency with which a term appears in a document — a term that occurs often is considered a good descriptor. One often removes "noise words" that are not meaningful, such as articles and conjunctions. It is also common to discount the weights of terms that occur in many documents, as they have low discriminating power. Then we can compute the similarity between two documents d_1 and d_2 by measuring the *cosine* between their vectors:

$$\cos(\vec{d_1}, \vec{d_2}) = \frac{\vec{d_1}}{||\vec{d_1}||} \cdot \frac{\vec{d_2}}{||\vec{d_2}||} = \frac{\sum_t w_{d_1,t} w_{d_2,t}}{\sqrt{\sum_t w_{d_1,t}^2} \sqrt{\sum_t w_{d_2,t}^2}}.$$

If the terms in d_1 are also present in d_2 and vice versa, the cosine is close to one; if the two documents do not share any terms, the cosine is zero. Note that the cosine is normalized by the norm, or size of each vector:

$$||\vec{d}|| = \sqrt{\sum_t w_{d,t}^2}.$$

This way, a longer document will not appear to be similar to many others just because it has many terms: its large norm would decrease the similarity.

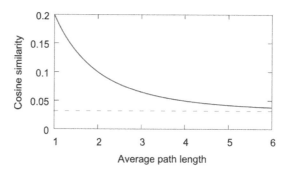

Web topical locality, measured by breadth-first search crawls from 100 sets of seed pages for as many topics. For each crawl, we report the average cosine similarity between seeds and pages as a function of the average path length between them. We observe a strong topical locality: the similarity between seeds and pages one link away is more than six times higher than the noise level expected between random pages (dashed line). As we crawl farther away, topic drift is illustrated by the decay in average similarity between seed and crawled pages, toward the noise level.

4.3 PageRank

We have learned that the pages retrieved by a Web crawler are processed by a search engine to build a search index — a data structure that lists all of the pages containing any given word or phrase. So, when you submit a query, the search engine can quickly list all the matching pages. But there may be millions of Web pages matching a certain query, say "social network," and you only have time to look at a handful. *Ranking algorithms* are therefore an equally critical component of a search engine. If results were sorted solely according to similarity between page content and query, users would have to look at a lot of low-quality pages, and even spam. But if results are ranked also taking some page importance or prestige measure into account, then the top results are more likely to be relevant, interesting, and reliable. Search engines began employing network centrality measures as ranking criteria in 1998, when Sergey Brin and Larry Page introduced *PageRank* as a component of a new search engine they called *Google.*

PageRank is an algorithm, or procedure, to compute a centrality measure that aims to capture the prestige or importance of each node; it is typically used in directed networks. It is also the name we give to the centrality measure itself. So when applied to the Web, the algorithm assigns each page a PageRank value. The ranking algorithm of a search engine can then use this value, in combination with many other factors — such as the match between query and page text — to sort the results of a query. A page with high PageRank is considered prestigious or important, and is given a boost by the ranking algorithm: other things being equal, pages with larger PageRank are ranked higher. Consider as an example a spammer who copies the content of the Wikipedia article on "social network" into a blog page filled with ads. The original and plagiarized page might look very similar in terms of content, but the Wikipedia page has a much higher PageRank. So when you submit the query "social network," you will see the Wikipedia article among the top results, while the spammer's page will be buried down and you probably will not see it.

The intuition for PageRank comes from imagining people who surf the Web at random, that is, following random links from page to page. This process is known as a *random walk* (or random surf) on the Web graph. The random walk is a simple model of user browsing or searching behavior; not knowing where a desired piece of information may be, or which link will lead to it, we make the simplest assumption that each link in a page has an equal chance of being clicked. We also wish to capture another user behavior, namely, that a user may at any time get tired of browsing and start a new browsing session. PageRank models this through occasional *jumps* from a current page to some other, randomly selected page among all pages in the network. The random jump process is called *teleportation.*

By imagining many people carrying out these modified random walk processes (surfing plus jumping) for a long time, one can measure how often each page would be visited. The fraction of times we land on a page is what we call the *PageRank* of that page. Figure 4.8 illustrates the PageRank values in a directed network; nodes with many paths leading to them are visited more often by random surfers, and consequently have high

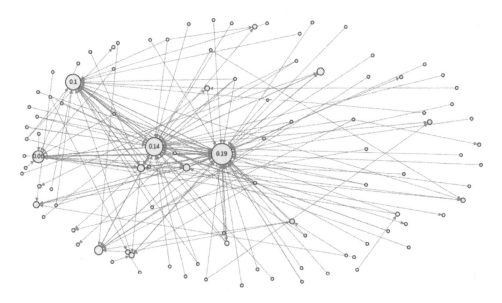

Fig. 4.8 PageRank in a directed network. The size of the nodes is proportional to their PageRank value, which is shown for a few of the nodes.

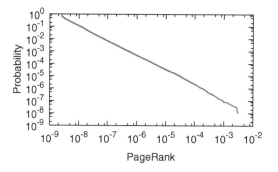

Fig. 4.9 Cumulative PageRank distribution of the Web host graph. PageRank values are normalized so that they sum to one. The network is obtained from the same 2012 crawl data used for Figure 4.5.

PageRank. In practice, PageRank is computed in a more efficient way, as explained in Box 4.2. Appendix B.1 presents an interactive demonstration of PageRank.

It turns out that the distribution of PageRank is quite similar to the distribution of in-degree on the Web (Figure 4.9). Why not just use in-degree for ranking then? To answer this question, consider that not all paths are equal. Paths from pages that are themselves visited often provide a larger boost. In other words, the importance of a page is affected by the importance of the pages that link to it — would you rather have a link to your homepage from your friend's blog, or from the front page of the *New York Times?* This is an important way in which PageRank differs from in-degree. Among two pages with the same in-degree, the one linked by pages with higher PageRank wins the game.

Box 4.2

PageRank

PageRank is usually computed from the link structure of a Web graph with an iterative approach called the *power method*. The idea is to compute how PageRank flows from page to page. The PageRank R of each node is initialized to some value (say $R_0 = 1/N$, so that all values sum to one). At each step, each node's value is refined, until the process converges, that is, none of the PageRank values change from one step to the next. We assume random jumps occur with probability expressed by a parameter α, called the *teleportation factor*, typically set to a small value $\alpha \approx 0.15$. The probability $1 - \alpha$, also called the *damping factor*, is instead associated with the random walk process. According to the PageRank model, at every step, with probability α a user jumps to a node selected at random among all pages; with probability $1 - \alpha$ the user continues browsing, following a random link from the current page. Therefore the PageRank of node i at time t is the sum of two terms that express the two ways in which one can arrive at page i:

$$R_t(i) = \frac{\alpha}{N} + (1 - \alpha) \sum_{j \in \text{pred}(i)} \frac{R_{t-1}(j)}{k_{out}(j)}. \tag{4.1}$$

The first term describes teleportation to i, which is one of N possible targets of a jump. The second term describes how one can traverse one of the links to i during the random walk: the sum is over the set of predecessors of i (i.e. the pages that link to i). Each of these pages, j, has $k_{out}(j)$ outgoing links. The PageRank of j is spread equally along these outgoing links, one of which leads to i. The figure below illustrates one step in the spreading process described by the second term of the equation ($\alpha = 0$) for page 3.

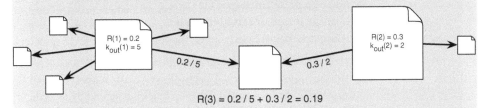

Note the *recursive* definition: the PageRank of a page depends on that of its neighbors. PageRank is a conserved quantity ($\sum_i R(i) = 1$), as it flows from one page to its neighbors through links, so it cannot be created or destroyed. For $\alpha > 0$ it turns out that, because teleportation connects all nodes virtually, PageRank is guaranteed to converge, and it does so relatively quickly — in less than 100 steps or so, even in very large networks. Therefore the power method is a much more efficient way to compute PageRank, compared to simulating modified random walks.

Of course, people play this game. The difference between a high and a low PageRank may mean appearing on the first page of search results or not, and that in turn may mean the survival or failure of a business. Many businesses thrive entirely thanks to their good search rankings. Therefore it should come as no surprise that there is an entire *search engine optimization* (SEO) industry to help websites improve their search rankings. Most SEO firms employ methods sanctioned by search engines, such as adopting descriptive page text and improving website navigability. However, less scrupulous SEO agents may employ methods of which search engines disapprove. These are often referred to as "spamdexing," or spamming the search index. One frequent spamdexing approach is that of creating *link farms*, large sets of fictitious websites that link to each other and to a target website. Such a clique structure is designed to trick PageRank-like algorithms and boost the target page's ranking. Search engines employ sophisticated network algorithms to combat link farms and other spamdexing attacks. When they detect this kind of abuse, they may remove a website from the search index.

NetworkX provides a function that runs the PageRank algorithm on a given directed network and returns a dictionary with the PageRank values of the nodes:

```
PR_dict = nx.pagerank(D)    # D is a DiGraph
```

4.4 Weighted Networks

So far we have focused on unweighted networks, in which links have a binary nature: either two nodes are linked, or not. However, connections between real-world entities are rarely so black-and-white. Very often links have attributes that allow us to compare two connections and determine which is stronger. In Chapter 0 we have seen some examples of weighted networks: the Twitter retweet network, where two accounts may retweet each other any number of times; the email network, where users can send any number of messages to each other; the Internet, where the data that travels on a physical link between two routers is measured by numbers of packets or bits; brain networks, where synapses between neurons transmit electrical firing signals at different rates; and food webs, where we can characterize the biomass of a prey species consumed by a predator species.

Even when we think of a network as unweighted, that is often just a reflection of our attempt to simplify how we model and analyze the underlying real-world relationships. Take the Facebook network, for example: we typically think of a friendship link as a binary relationship. However, not all friendships are equal; two close friends may have many common contacts, may like and comment on each other's posts many times, and may reshare and tag each other's photos. The same is not true for two remote acquaintances. The unweighted Facebook friend network is just a very simplified model of actual relationships. But make no mistake: platforms like Facebook monitor your every action and are well aware of the strength of each of your links. Other social networks are similar; the movie co-star network, for instance, can have weighted links based on the number of movies in which two stars have acted together. In Box 5.6 we will see that the strength of social ties has long played an important role in the study of social networks.

Information and transportation networks provide more examples of weighted, directed networks: in the Web and Wikipedia, some links are clicked much more often than others. And in air traffic networks, some connections carry more flights and passengers than others.

In all of these networks, we use link *weights* to represent important measures such as messages, bits, likes, clicks, and passengers. Recall from Chapter 1 that the degree, in-degree, and out-degree node centrality measures are extended to the *strength, in-strength*, and *out-strength* in weighted networks.

4.5 Information and Misinformation

Let us delve a bit deeper into the features of weighted directed networks, using *information diffusion networks* as a case study. In such networks, nodes represent people and links represent pieces of information — ideas, concepts, news, or behaviors — that are passed from person to person. A transmissible unit of information is called a *meme* — images with captions are only one kind of Internet meme. Twitter provides us with data that is ideal to observe how images, movies, Web links, phrases, hashtags, and other memes spread online. Each of these memes is uniquely identified by a text string: a URL for a Web link or media entity, or a label prefixed by the hash mark (#) for a hashtag that expresses a concept or topic. A tweet may carry multiple memes. For example, the message *"Hoosiers are the best #GOIU iuhoosiers.com"* contains the hashtag `#GOIU` and the Web link `https://iuhoosiers.com/`.

Using data from Twitter, we can build various kinds of diffusion networks capturing different ways in which a meme can spread: via retweets, quoted tweets, mentions, and replies.[1] For example, if Alice mentions Bob in a tweet, Bob is likely to see the tweet and therefore we can assume that the tweet has spread from Alice to Bob. Similarly, if Alice replies to Bob, we can infer that the original message has been delivered from Bob to Alice. For simplicity, let us focus on retweets. If Alice follows Bob, then Alice may retweet a message from Bob and in so doing spread a meme contained in the tweet to all of her followers. A *retweet cascade* is a directed tree that captures how a meme propagates from its originator to all the users who are eventually exposed to it. However, as illustrated in Figure 4.10, it is not straightforward to reconstruct a cascade tree from Twitter data. Still, we can easily observe all the users who are exposed to the meme. They are all connected to the origin, forming a star network. If the underlying follower network is known, it is possible to reconstruct an approximation of the cascade tree.

The same meme can generate many cascade trees (stars). For example, multiple users can share the same link to a news article, or tweet using the same hashtag. By aggregating all of these star trees, we obtain a *forest* (set of trees) that we call the diffusion network.

In building a diffusion network, one must first formulate the meme, or memes, whose spread we want to analyze. We might be interested in a single hashtag, say `#elections` or `#soccer`; or in all links to articles from a news source or a set of news sources. For instance, when investigating online misinformation, we may wish to track stories from low-credibility sources that routinely spread fabricated news, hoaxes, conspiracy theories, hyper-partisan content, click bait, or junk science.

---

[1] You can explore interactive diffusion networks from Twitter using a tool from the Observatory on Social Media at osome.iuni.iu.edu/tools/networks/. You can also generate animation movies showing how these networks unfold over time, using another tool at osome.iuni.iu.edu/tools/movies/.

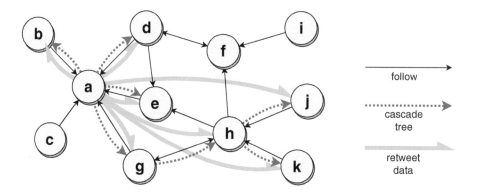

Fig. 4.10 Illustration of follower and retweet networks on Twitter. Users typically see, and sometimes retweet, the tweets of accounts they follow. In this example, **b** follows **a** and also retweets a message from **a**. **c** also follows **a**, but does not retweet **a**. A user may also retweet a message that they see because it was retweeted by someone they follow. Here **h** does not follow **a**, but retweets a message that originated from **a** and that was retweeted by **g**, who is followed by **h**. By tracing these retweet chains, one could in theory reconstruct the *retweet cascade tree* with **a** (the origin of the tweet) as root node. However, Twitter does not provide data about retweet cascades. Instead, each retweet points directly to the origin of the tweet. Therefore the cascade tree becomes a star network with the same root node.

We can think of the cascade forest associated with each hashtag as a layer in a multilayer network, as discussed in Section 1.8. And since retweets occur at different times, this is also a temporal network. The diffusion network in Figure 0.3 was obtained by aggregating cascade forests across time and across many popular hashtags associated with political conversations during the 2010 US midterm election.

Once we build a diffusion network, we can observe several features. Is the spread uniform across the network, or concentrated within dense and segregated clusters of nodes? For example, Figure 0.3 shows a polarized structure with two mostly segregated communities: conservatives and progressives, each retweeting almost exclusively members of the same group. These communities are sometimes called *echo chambers*, because a user is mostly exposed to opinions reinforcing their own. Figure 4.11 shows another example of an echo chamber, made up of people who are vulnerable to political misinformation. Observe that the users who spread the misinformation share very few articles from fact-checking sources. In Chapter 6 we will learn how to detect such communities in networks.

Another property of interest is the *virality* of a meme. The simplest way to quantify this is to measure the size of the diffusion network (i.e. the number of users exposed to the meme). However, given a large network, its structure is also revealing. For instance, a star network where many followers of a celebrity retweet a meme may reflect the popularity of the celebrity more than that of the meme. However, assuming we can reconstruct cascade trees, a deep network with long chains of retweets may indicate wider appeal of the message. As illustrated in Figure 4.12, misinformation is often more viral than actual news reports.

Diffusion networks can provide us with insight about patterns of information production and consumption. In a meme's weighted and directed retweet network, the weight of a link from Alice to Bob indicates how many times Bob has propagated the meme originating

Retweet subnetwork for articles during the run-up to the 2016 US election. Each of the $N = 52{,}452$ nodes represents a Twitter account and each of the links represents a retweet linking to an article from a low-credibility (purple) or fact-checking (orange) source. The visualized subnetwork is the $k = 5$ core of the full retweet network (Section 3.6). Image adapted from Shao *et al.* (2018a) under CC BY 4.0 license.

from Alice's posts. We can think of Alice as a producer and Bob as a consumer of information about the meme. Extending this analysis to the entire network, we can use the out-strength and in-strength of a node to measure the user's propensity to produce and consume information, respectively — to be retweeted by others or to retweet others. A user may play both roles, of course. Therefore we can look at the ratio between out-strength and in-strength to classify a user: if the ratio is much larger than one, the user is mainly a producer; conversely, a ratio below one indicates a consumer.

Whether we focus on one meme or aggregate across all messages, a high value of out-strength can also be used as an indicator of *influence*, in the sense that messages from the user get retweeted a lot. Figure 4.12 and the image on the book's cover both illustrate influential nodes in diffusion networks by drawing nodes with size proportional to their out-strength. Using out-strength as a proxy for influence is in contrast to the in-degree of the follower network (i.e. the number of followers), which measures popularity but not necessarily influence; one could have many followers who do not retweet. By comparing these two quantities — number of retweets versus number of followers — we can get a better sense of one's influence.

Given the huge power of social media to inform, persuade, and influence all of us, it should come as no surprise that increasing resources are being devoted to manipulate these platforms. It is pretty easy and cheap to buy fake followers to boost an account's perceived popularity on Twitter. This amounts to adding nodes and links to increase a node's in-degree in the follower network, similarly to how one might create fake websites and links to boost a site's PageRank. Diffusion networks are also manipulated through *social bots*, deceptive fake accounts that impersonate users. Bots can be used to generate fake tweets and create the appearance of grassroots campaigns, or *astroturf*. In this way they

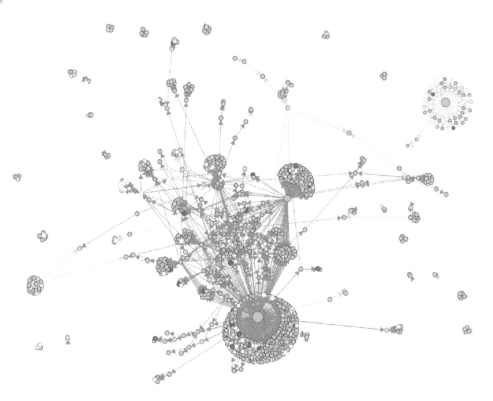

Fig. 4.12 Retweet network for two articles about the White Helmets, a volunteer rescue organization acting during the civil war in Syria. The White Helmets have been the target of a disinformation campaign, with false claims of terrorist ties and other conspiracy theories. The gray and yellow links depict the diffusion of one of these false claims and a fact-checking article, respectively. We can easily observe that the spread of the misinformation is more viral. Node size is proportional to out-strength and node color captures the probability that an account is automated: blue nodes are likely human and red nodes are likely bots. Image from Hoaxy, a tool that visualizes the spread of misinformation on Twitter (hoaxy.iuni.iu.edu).

can trick both human users and ranking algorithms, effectively hijacking public attention. Bots can also be used to retweet some messages, amplifying their perceived engagement and boosting their spread. Figure 4.12 and the image on the book's cover illustrate how bots are used to amplify the spread of misinformation and manipulate public debates. In fact, bots can gain significant influence. Figure 4.13 illustrates another form of network manipulation leveraging fake replies and mentions. In this way one can target influential and popular users, such as reporters and politicians, and expose them to misinformation in the hope that they will spread it to their many followers. Yet another misinformation diffusion network manipulated by influential bots can be seen in Figure 7.1 (Chapter 7).[2] While our examples are focused on Twitter, other social media platforms like Facebook, Instagram, and WhatsApp are also exploited for spreading misinformation.

---

[2] You can explore the role of bots in interactive diffusion networks from Twitter using the Hoaxy tool from the Observatory on Social Media at hoaxy.iuni.iu.edu.

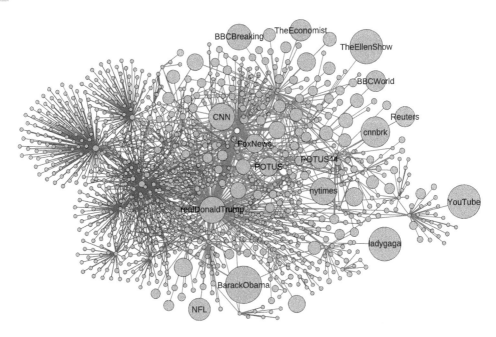

Fig. 4.13 A portion of the diffusion network for a fake news report that claimed massive voter fraud by illegal aliens in the 2016 US election. Despite presenting no evidence and being debunked by fact checkers, the article was shared over 18,000 times on Twitter. In this visualization, node size represents the number of followers of an account; links illustrate how the article spread by retweets or quoted tweets (blue), and by replies or mentions (red); and the width of a link represents its weight — number of retweets, quotes, replies, and mentions between two accounts. The small yellow node near the center is a bot that systematically tweets the misinformation in replies to messages mentioning the US President. The resulting mentions generate a thick red link connecting the bot and @realDonaldTrump. Image from Shao *et al.* (2018b) under CC BY 4.0 license.

4.6 Co-occurrence Networks

We have already discussed several examples of weighted networks in this chapter. Information, communication, transportation, biological, and even social networks often have weighted links. Another way in which weighted networks can arise is via relationships between more than one type of entity.

The simplest case is a directed network, in which each link has a source and a target node. Imagine moving all source nodes to one side and all target nodes to the other (nodes that are both sources and targets can be duplicated and appear on both sides). To use a concrete case, refer to the citation network in Figure 4.1. A citation link is a relationship connecting two different types of entities, the citing paper (source) and the cited paper (target). We can now construct a new network among papers in each of these groups. Two papers are *co-cited* if there are one or more papers that cite both of them. Their co-citation count is the number of papers citing both. Similarly, two papers are *co-referenced* if there are one or more papers cited by both of them. Their co-reference count is the number of papers cited by both. *Co-citation* and *co-reference* networks are undirected, weighted networks where the links are weighted by co-citation and co-reference counts, respectively. They are often used to find related sets of publications.

There are many situations in which relationships between two distinct types of entities are naturally represented by *bipartite networks* — where each link connects two nodes of different types. One example that we discussed in Chapter 0 is the relationship between actors/actresses and movies in which they starred. As shown in Figure 0.2, we can create from this bipartite network a network between actors/actresses who have co-starred in movies. The links in the co-star network can be weighted by the number of movies in which two people have co-starred. Generating weighted networks from bipartite networks in this way, called *projection*, is common practice, and the resulting weighted networks are called *co-occurrence networks* because links represent two entities of one type that "occur" together in association with one or more entities of another type. Other common examples of co-occurrence networks include students who take the same classes, products such as movies and books purchased by the same customers, and pages co-liked by Facebook users. Each time you "like" or share something on a social media platform, you create a link between you and the object [Figure 4.14(a)]. These links are shared with friends, but also aggregated by the platform across millions of people, yielding massive co-occurrence networks [Figure 4.14(b)] that can be used for recommendation and targeted advertising.

A bipartite network can of course have weighted links. Rating systems have weights that represent, say, how much one enjoys a movie or a mobile app. Another source of weighted bipartite networks is *social tagging*: a person annotates a resource (identified by a URL) with one or more labels, or *tags*. Sharing sites like Flickr.com and YouTube.com popularized social tagging for images, movies, and other media. The elementary construct of the social tagging representation is the *triple* (u, r, t), where user u tags resource r with tag t. A resource can be a media object, scientific publication, website, news article, and so on. Tagging can be implicit in social media. For example, many Twitter users link to

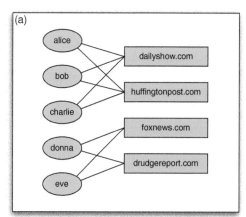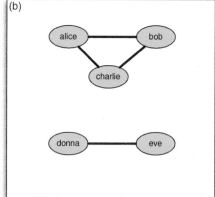

Fig. 4.14 (a) A bipartite network induced by "like" relationships. (b) A user co-occurrence network derived from projecting the "like" network onto user nodes.

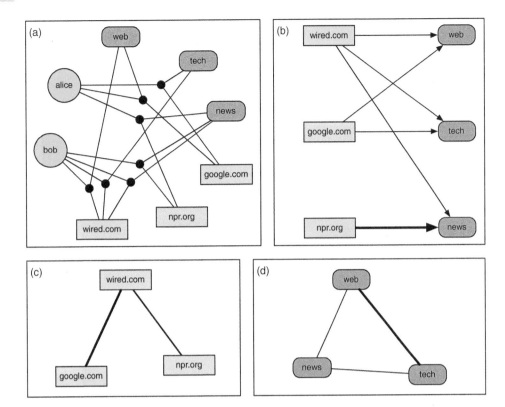

Fig. 4.15 Example of a folksonomy and derived bipartite and co-occurrence networks. (a) Two users (alice and bob) annotate three resources (npr.org, wired.com, google.com) using three tags (news, web, tech), resulting in seven triples. (b) Projecting the folksonomy onto resources and tags, we obtain a bipartite network. Link weights correspond to numbers of triples, or numbers of users. The link from npr.org to news has a larger weight because both users agree on that annotation. (c) A resource co-occurrence network. The resources wired.com and google.com are more similar because they co-occur with two tags, web and tech. (d) A tag co-occurrence network. The link between web and tech has a larger weight because of the similarity in their resources: the two tags co-occur with two resources, wired.com and google.com.

news articles or blog entries while also labeling their posts with hashtags. Triples can be extracted from such tweets — one for each pair of link and hashtag, with the author of the post as user. Figure 4.15(a) illustrates a set of triples. When aggregated across many users, such a set is called a *folksonomy*, because it is a taxonomy that emerges from many people. Folksonomies can be useful for searching or suggesting websites.

From a folksonomy, we can extract a bipartite network by projecting the triples onto two of the node types. The resulting links connect only nodes of one type (say, tags) to nodes of another type (say, resources). We can therefore think of these links as directed, as shown in Figure 4.15(b). The links can also have weights representing, say, how many users have annotated a particular resource with a particular tag [Figure 4.15(b)]. This way, instead of losing the information about the users, we have encoded it into a measure of link reliability.

From a bipartite network, as discussed earlier, we can create co-occurrence networks by further projecting one type of node based on shared neighbors of the other type. For example, in Figure 4.15(c) we project onto a resource network and in Figure 4.15(d) we project onto a tag network.[3] The direction of the links is lost in these co-occurrence networks, but we can preserve weight information by comparing how two tags are connected to the resources.

One way is to represent, say, a tag as a vector of resources $\vec{t} = \{w_{t,1}, \ldots, w_{t,n_r}\}$, where $w_{t,r}$ is the number of people tagging resource r with tag t, and n_r is the total number of resources. So each element of the tag vector is a weight representing the association of a resource with the tag: the weights of the links in Figure 4.15(b). Then we can compute the *cosine similarity* between two tag vectors (see Box 4.1) and use it as a co-occurrence link weight. If the two tags are used to annotate resources in similar ways, the weight is high; if they never co-occur, the weight is zero and the tag nodes are not linked.

4.7 Weight Heterogeneity

In a weighted network, the weights of the links may carry important information about the process or relationships modeled by the network. Links with different weights may represent very different associations. To explore this difference, consider another class of weighted networks: those that capture various kinds of traffic. Traffic networks include air and other transportation networks, where weights represent passengers or flights between airports, or cars between intersections; the Internet, where weights denote packets or bits of data between routers; and Wikipedia, with weights representing clicks between articles. Let us focus on the case of Web traffic, similar to Wikipedia but extended to all websites.

4.7.1 Web Traffic

Data about Web traffic can be captured by browsers (or browser toolbars, or extensions) that record click data and transmit it to a collection server. Alternatively, an ISP may monitor packets that carry HTTP/HTTPS requests, which include the target host and page as well as the source URL, called *referer*. Both of these collection methods are somewhat biased; in the former case we only observe traffic generated by users of browsers instrumented with traffic-monitoring software, in the latter we can only see packets that pass through the ISP routers. Nevertheless, both methods enable very large collections of Web traffic data. One can count clicks between individual pages, or, as we do next, consider the aggregate traffic at the level of entire websites identified by their hostnames, such as en.wikipedia.org, google.com and www.indiana.edu.

---

[3] You can explore interactive hashtag co-occurrence networks from Twitter using a tool from the Observatory on Social Media at osome.iuni.iu.edu/tools/networks/.

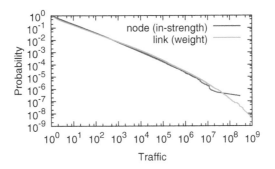

Fig. 4.16 Cumulative distributions of node in-strength (website traffic) and link weight in a Web traffic network. Almost one billion clicks, representing the combined Web browsing activities of about 100,000 anonymous users, were collected at Indiana University between 2006 and 2007. The resulting network has about 4 million sites and 11 million weighted directed links.

Inspecting Web traffic networks allows us to study the distribution of website traffic (total number of clicks to a site), expressed by node in-strength, and of link traffic (total number of clicks on a hyperlink), expressed by link weight. The heavy tails in Figure 4.16 reveal that both of these distributions are extremely heterogeneous. Most websites receive very few clicks, while some receive massive traffic. Similarly, many links are hardly ever clicked, while others are clicked all the time.

Recall from Section 4.3 that the idea of PageRank was to simulate Web users who browse the Web. By comparing the ranking of network nodes according to their in-strength with the ranking produced by PageRank, we can ask whether PageRank is able to predict traffic from the structure of the Web link graph. In other words, does the random surfer model capture the aggregate browsing patterns of actual Web users? As it turns out, the answer is no: despite similar heavy-tailed distributions (see Figures 4.9 and 4.16), the correlation between PageRank and traffic is quite weak. Therefore, some of the simplifying assumption of the PageRank model must be violated by our Web browsing behavior.

To get an idea of which ingredients of the random surfer are least realistic, let us consider the ratio between out-strength and in-strength of nodes. Except for the start and end nodes of browsing sessions, the ratio must be one because the flow of traffic that goes into a node equals the flow that comes out of the node. According to PageRank, teleportation does not favor any nodes; each node is equally likely to be the target of a random jump (where browsing begins) and equally likely to be the source of a random jump (where browsing ends). Therefore, even with teleportation, the PageRank model produces a ratio between out-strength and in-strength very close to one for all nodes. We would then expect a narrow distribution peaked around one. But Figure 4.17 paints a very different picture, with huge fluctuations spanning many orders of magnitude: some nodes are much more likely to be starting points and others are much more likely to be ending points of browsing sessions. This is not surprising; we tend to start surfing from a few familiar, bookmarked sites. However, most sites are not very interesting, and are therefore more likely to be places where we stop and jump away. We conclude that random teleportation is an unrealistic ingredient of the PageRank model.

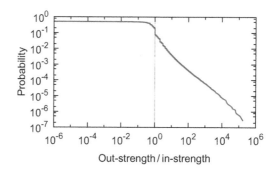

Fig. 4.17 Cumulative distributions of the ratio between node out-strength and in-strength in the Web traffic network described in Figure 4.16. Strength ratios $s_{out}/s_{in} \ll 1$ indicate sites where browsing sessions are more likely to terminate, while $s_{out}/s_{in} \gg 1$ for starting points.

4.7.2 Link Filtering

Dense networks are hard to visualize and study: they look like "hairballs" and many links are not significant. For these reasons, it is often helpful to prune low-weight links in a weighted network. This is especially needed in co-occurrence networks, were many links with low weight may result from noise. As an example, consider a network of documents or Web pages where links are defined by text similarity — shared keywords from the textual content of the nodes. A pair of unrelated or weakly related documents are likely to be linked because they share a few generic keywords. In cases like this, we need a method to filter such links and obtain a sparser network with only the meaningful connections.

The simplest approach to prune a network is to remove all links with weight below some threshold. This method works well in many settings. However, there are also many cases in which filtering based on a global threshold does not work. To see why, consider a network with a heavy-tailed distribution of link weights — a pretty common scenario in co-occurrence, traffic, and other weighted networks (recall Figure 4.16). The weights are so heterogeneous that it is impossible to find a good threshold: we are likely to preserve some links that are not significant, and/or disconnect many nodes with low strength. For low-strength nodes, low-weight links may be significant, even though links with the same weight may be insignificant for nodes with much higher strength.

To get around this problem, we need to use different thresholds for different nodes. One way to accomplish this would be to define a threshold relative to the degree or strength of each node: we might keep only the 10% of links with largest weight for each node, or the links with largest weight that account for 80% of a node's strength. But even so, we cannot be sure that we are retaining all of the significant links, or whether we are retaining some that are not significant. A more principled approach is to find the *network backbone*, (i.e. detect the links that carry a disproportionate fraction of each node's strength). These are the most significant links to be preserved. Box 4.3 describes how this can be done. Figure 4.18 illustrates a backbone network extracted from a dense network.

Network Backbone

In networks with broad distributions of link weights, using a global threshold to prune links is inappropriate. Instead, we can use the weight fluctuations for each node to identify the links to be preserved — those that carry most of the weight. Given a node i with degree k_i and strength s_i, let us evaluate a link against a null model in which the weights are distributed randomly on the k_i links adjacent to i, with the constraint that their sum equals s_i. The probability that a link has weight w_{ij} or larger under this hypothesis is

$$p_{ij} = \left(1 - \frac{w_{ij}}{s_i}\right)^{k_i - 1}. \qquad (4.2)$$

So, if link ij has weight w_{ij}, from Eq. (4.2) we compute the probability p_{ij} that such a value is compatible with the null model: if $p_{ij} < \alpha$, where α is a parameter that represents the desired significance level, the link is preserved, otherwise it is removed. Lower values of α lead to sparser networks, as fewer links are preserved. Since a link is connected to two nodes, we can obtain two values for p_{ij} by plugging the strength and degree of either node into Eq. (4.2). We can then use the larger or smaller of these values, depending on whether we wish to prune more or less aggressively. This link filtering procedure extracts a *network backbone*, which is supposed to preserve the essential structure and global properties of the network.

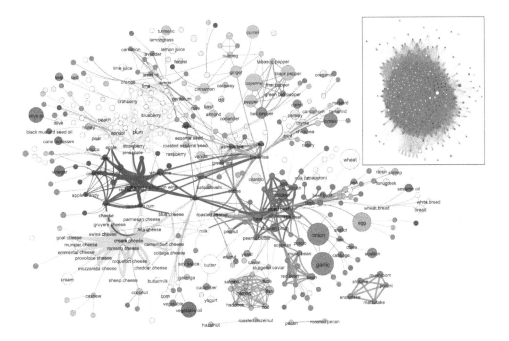

A flavor network: each node denotes an ingredient, its color indicates a food category, and its size reflects the ingredient prevalence in recipes. Two ingredients are connected if they share flavor compounds, with the link width representing the number of shared compounds. The full network is shown in the inset, while the main image visualizes the backbone network with the significant links identified by the method in Box 4.3, using $\alpha = 0.04$. Images adapted from Ahn *et al.* (2011) under CC BY 4.0 license.

4.8 Summary

Information networks, such as Wikipedia and the Web in general, have directed links. So do many biological networks, such as brains; communication networks, including email and the Internet; transportation networks, such as air flights; and some social media, notably Twitter. Links are often weighted to represent, for example, intensity of interaction or similarity between nodes. Network weights can also be used to represent traffic flowing between nodes: clicks, messages, packets, travelers, retweet counts, and so on. We explored features of directed and weighted networks by focusing on several cases:

1. The Web forms a huge information network, with a virtually infinite number of pages connected by hyperlinks. Browsers use the HTTP protocol to navigate links and download page content. This content is usually expressed in the HTML language, that specifies how rich content — text and embedded media — is presented.
2. We study the structure of the Web and host graphs, where each node represents a page or website, respectively, using data collected by Web crawlers — programs that browse the Web automatically and allow us to reconstruct large samples of the network. The Web has a heavy-tailed distribution of in-degree and ultra-short paths facilitated by hugely popular hub pages. It also has a very large strongly connected component within the giant component.
3. We can represent documents, such as Web pages, as high-dimensional vectors of words, and use the cosine between these vectors to measure text similarity among pages. In this way we can study topical locality, the relationship between network connections and page content. Because authors tend to link related pages, the Web has a clustered structure such that pages within a short distance of each other in the network are likely to be similar and semantically related.
4. PageRank is a node centrality measure based on a random walk model of Web surfing, modified with random jumps. While people in reality do not browse the Web in such a random manner, PageRank is usually applied to measure the prestige of Web pages. The PageRank algorithm works in any directed network, but is especially important for its role in ranking search engine results. It was the key ingredient of Google when it was introduced.
5. Information diffusion networks arise when we share content on social media, for example by retweeting links, images, and hashtags. The resulting cascade networks allow us to track the spread of news, ideas, beliefs, and even misinformation. The size and structure of these graphs help us identify viral concepts. Using node out-strength and in-strength, we can characterize the roles of people who produce and consume information. Nodes with high strength, especially relative to their degree, flag accounts that are more active or influential. Social bots can be used to manipulate these networks.
6. Weighted networks often result from bipartite graphs. The weight of a link between two nodes **a** and **b** of the same type is a measure of how many nodes of another type are associated with both **a** and **b**. Such networks originate from co-occurrence relations like co-citations/co-references, product recommendations, and word/tag similarity.

7. Weighted networks, such as those obtained from traffic and co-occurrence data, can be very dense. Therefore it is often necessary to prune the graph by filtering out the low-weight links. However, weighted networks often have heavy-tailed weight distributions. In these cases, using a global weight threshold isolates the majority of the nodes. By determining a local weight threshold to identify the statistically significant links for each node, we can extract the backbone of a heterogeneous weighted network.

4.9 Further Reading

You can read about the vision, design, and history of the Web in a book coauthored by its inventor (Berners-Lee and Fischetti, 2000). To learn more about how search engines work, consider the textbooks on information retrieval by Baeza-Yates and Ribeiro-Neto (2011) or Manning *et al.* (2008). Liu (2011) goes into how to mine data from Web link, content, and usage networks; its chapter 8 focuses on Web crawlers.

Albert *et al.* (1999) first analyzed the average path length of the Web in 1999, based on a crawl of Notre Dame University websites. The Web was thought to contain about a billion pages back then, so the authors extrapolated from a logarithmic fit between average path length and (sub)network size to estimate that the Web had a diameter of 19 links. The following year, Broder *et al.* (2000) reported on the first systematic study of the Web structure. They measured the average path length on a much larger Web crawl with about $N = 10^8$ pages, in rough agreement with the earlier prediction. A more recent analysis of a much larger Web crawl was carried out by Meusel *et al.* (2015).

Barabási and Albert (1999) reported the first evidence of the heavy-tailed distribution of Web page in-degree. Broder *et al.* (2000) later confirmed this from a larger crawl. Broder *et al.* (2000) also analyzed the bow-tie structure of the directed Web graph. Serrano *et al.* (2007) showed that the relative sizes of the largest strongly connected component and in/out-components depend on the particular crawler used to reconstruct the Web graph.

Davison (2000) measured topical locality on the Web by comparing the content of pairs of pages selected at random, linked by a common predecessor (siblings), and connected by a hyperlink. Menczer (2004) extended this analysis by performing a breadth-first crawl to consider how content and semantic similarity decay for pages within a certain distance of each other (Figure 4.7).

The idea of using network centrality measures to rank search engine results was conceived by Marchiori (1997). A year later, Brin and Page (1998) introduced Google and described how PageRank was used to rank its search results. It turns out that the same measure of centrality had been proposed 50 years earlier by Seeley (1949) as a way to gauge the importance of a person in a social network. A related authority measure, based on a bipartite representation of the Web graph, was proposed by Kleinberg (1999). Fortunato *et al.* (2007) showed that the average PageRank score of nodes with equal in-degree is proportional to the in-degree. For a review of the math behind PageRank, see Gleich (2015).

Dawkins (2016) introduced the concept of *meme* to mean a unit of information, belief, or behavior that can be transmitted from person to person. This was a precursor of the images, hashtags, and links that now spread on social media. Goel *et al.* (2015) proposed a definition of structural virality for memes and a method to reconstruct retweet cascade networks on Twitter. Studying these diffusion networks, Cha *et al.* (2010) showed that having a high degree (many followers) is not the only factor that affects a node's influence.

A strong polarization of communication networks on Twitter was observed by analyzing the diffusion networks of political hashtags, with segregated conservative versus progressive communities (Conover *et al.*, 2011b). Similarly, Shao *et al.* (2018a) find segregated communities sharing misinformation versus fact-checking articles. A social network community with homogeneous opinions, isolated from different views, is sometimes referred to as an echo chamber (Sunstein, 2001) or filter bubble (Pariser, 2011).

Ratkiewicz *et al.* (2011) observed the earliest instances of fake news websites producing misinformation to be spread via social media. The factors that affect the viral spread of misinformation in social networks are the subject of extensive investigation (Lazer *et al.*, 2018); they include novelty (Vosoughi *et al.*, 2018) and amplification by social bots (Ferrara *et al.*, 2016; Shao *et al.*, 2018b).

Meiss *et al.* (2008) collected massive data about Web clicks to reconstruct a large Web traffic network, revealing the limitations of PageRank as a model of Web surfing. Better models account for bookmarking of popular starting nodes, backtracking (or browser tabs), and topical locality (Meiss *et al.*, 2010). Meiss *et al.* (2008) also showed that the distribution of link weights has a heavy tail. Serrano *et al.* (2009) introduced a method to extract the backbone of networks with heterogeneous weights.

Exercises

4.1 Go to scholar.google.com and search for publications on a topic that interests you. Pick two papers from the list of search results.
1. What is the in-degree of each of the two papers in the citation network?
2. For each of the two papers, look at the lists of other papers citing them (click on "Cited by…"). Calculate the co-citation between the two papers. (*Hint*: This can be tedious if you select two papers with too many citations.)
3. What is the out-degree of each of the two papers in the citation network? (*Hint*: The papers must be available for access or download to answer this question.)
4. Download the two papers and analyze the lists of references. Calculate the co-reference (also known as *bibliographic coupling*) between the two papers.

4.2 Go to the Wikipedia article on "network science" (en.wikipedia.org/wiki/Network_science).
1. What is the out-degree of this page in the Wikipedia network? (*Hint*: For simplicity, in this exercise you can focus on the outgoing links in the "See also" section,

which typically includes several links to other Wikipedia articles; if this section is not present, you may assume $k_{out} = 0$.)

2. Visit the successor nodes of the "network science" node in the Wikipedia network and report on how many of the outgoing links from this article are reciprocal.

3. Build the ego network of the "network science" node and find the largest strongly connected component. Recall that the ego network consists of one node (the ego), all of its neighbors, and all of the links among them (see Figure 2.8). The definition of a directed ego network is analogous, replacing neighbors by successors.

4. What is the node in the "network science" ego network with the maximum out-degree?

5. What is the node in the "network science" ego network with the maximum in-degree?

4.3 Consider the Wikipedia "network science" ego network constructed in the previous problem. Represent each of these nodes as a list of categories — these are found at the bottom of each Wikipedia article; an example is "Network theory." For each pair of nodes, calculate the cosine similarity between the category vectors. (*Hint*: A list is a vector where the weight of each category is one, and the weight of any category not in the list is zero.)

1. Which two articles from your sample are most similar to each other? What is the value of the cosine similarity?

2. Which two articles from your sample are least similar to each other? What is the value of the cosine similarity?

3. Do your measurements provide evidence of topical locality? Why or why not? (*Hint*: If you disregard link directions, any two nodes are either at distance one from each other, if connected, or at distance two, through the ego. Compare the average similarity of article pairs in these two groups.)

4.4 Consider the small network in Figure 4.19. Initialize the PageRank of each page with the value $R_0 = 1/3$. Apply Eq. (4.1) without teleportation ($\alpha = 0$) to calculate the values of PageRank in the next iteration ($t = 1$). Continue to update the values until convergence — assume the values have converged when there is no change in the third decimal digit of each node's PageRank. [*Hint*: Be sure to use the values from the previous iteration when computing the new values; for example, use the initial values ($t = 0$) when calculating the $t = 1$ values.] After how many iterations do the values converge? What are the final values of PageRank?

4.5 Repeat the previous exercise using the network in Figure 4.19, but this time use a teleportation parameter $\alpha = 0.2$. What is t at convergence and what are the PageRank values?

4.6 Go to the PageRank demo page at go.iu.edu/pagerank and insert a bunch of nodes (names), node text attributes (colors), and links. The demo calculates the PageRank

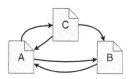

A directed network representing a small website with three pages and their hyperlinks.

and you can use it to measure the popularity of each node. Observe how the values change with the addition of new nodes and links.[4]

1. Who is the most popular? What can you do to increase your PageRank?
2. The search feature of the demo works like a super-simplified search engine. Search for some colors. How does the similarity between your query and a node's text attributes affect the ranking of the node? How does PageRank affect the ranking?

4.7 Download the Wikipedia dataset (`graphml` file) from the book's GitHub repository,[5] in the `enwiki_math` folder. Use NetworkX to load the file as a directed network (digraph), then run the PageRank algorithm to compute the PageRank of each article.

1. What are the top 10 articles by PageRank?
2. Compare the top 10 articles by PageRank with the top 10 articles by in-degree. Are they the same? Why or why not?

4.8 Go through the Chapter 4 Tutorial on the book's GitHub repository.

4.9 Test the Friendship Paradox (discussed in Chapter 3) on Twitter. Refer to the Chapter 4 Tutorial for usage of the Twitter API. Since Twitter is a directed network, we can formulate the Friendship Paradox in terms of in-degree/out-degree of a node's successors/predecessors. One version is to ask: "do your friends (the people you follow) have more followers than you, on average?" In this case we want to measure the in-degree of your successor nodes in the follower network. If you are not a Twitter user, you can answer this question using someone else's Twitter handle, such as @clayadavis.

1. Assuming `user` is the response from a query to `users/show.json` about @clayadavis, which of the following will give you the number of people followed by @clayadavis?
   a. `user['friends']['count']`
   b. `user['friends_count']`
   c. `user['followers']['count']`
   d. `user['followers_count']`

---

[4] *Note to the instructor:* This exercise is more fun in a large study group. Each participant can use their own laptop, use an anonymous pseudonym in place of their name, enter their favorite colors, and link to their friends. Or conduct the exercise in class, with extra credit assigned to the top students after a group activity. Tell your students to consider alliances and be ruthless!

[5] github.com/CambridgeUniversityPress/FirstCourseNetworkScience

2. Verify if the Friendship Paradox holds in your case by calculating the average number of followers of your friends, and comparing it with your number of followers. If your account has more than the maximum number of friends that can be returned by the Twitter API with a single request (200 at the time of writing), you will need more than one API call to obtain the full list of friends. Refer to the tutorial for using a cursor to get multiple pages of results. What is the average number of followers amongst all of your friends?

3. Mathematically, the Friendship Paradox makes a statement about averages: "On average, your friends have more followers than you do." A different statement would be "Most of your friends have more followers than you do." This second statement is about the median, not the average, and is in fact a stronger statement (less likely to be true). Is it true here? Do most of your friends have more followers than you do? To answer this question, measure the median number of followers amongst all of your friends.

4. Amongst your friends, what is the screen name of the user with the most followers?

4.10 Build the retweet network for `#RepealThe19th`, a controversial hashtag used during the 2016 US presidential campaign to advocate for the repeal of the 19th amendment to the US Constitution, which grants women the right to vote. The data for this exercise is provided in the datasets directory of the book's GitHub repository. The file named `repealthe19th.jsonl.gz` includes 23,343 tweets containing the hashtag. Each line is a tweet JSON object. After parsing the file, you should verify that you have this many tweets; a deviation from this number would flag parsing errors that may affect your answers. Refer to the Chapter 4 Tutorial for how to use these tweets to create a retweet network in order to answer the following questions. Keep the following in mind. (i) Link direction follows information flow: if Alice retweets Bob, there is a link from Bob to Alice. (ii) Remove self-loops; you can do so after creating the network, or modify your network creation code to not add them at all. This will definitely affect some answers.

1. How many nodes are in the retweet network?
2. How many links are in the retweet network?
3. What is the screen name of the node with highest out-strength in the network? What is its out-strength?
4. What is the screen name of the node with the second-highest out-strength in the network?
5. What is the screen name of the node with highest in-strength in the network? What is its in-strength?
6. Describe what the out-strength and in-strength values of these accounts tell you about their online behavior.
7. What is the ID of the most retweeted tweet? Use the `id_str` attribute; when working with JSON files, this is usually good practice due to tweet IDs being 64 bits. Note that, given a tweet ID, you can see the tweet by visiting

the following URL in your browser: https://twitter.com/user/status/<tweet-id>, replacing <tweet-id> with the numeric tweet ID.

8. How many nodes in the retweet network have zero out-strength?
9. Which of the following best describes what it means for a node to have zero out-strength in this network? Each of the statements applies only to the sample of tweets we used to build this network.
   a. The user produced no tweets
   b. The user hasn't retweeted anyone else
   c. The user hasn't been retweeted by anyone
   d. The user has no followers
   e. The user does not follow any other users
10. Which of the following best describes the connectedness of this network?
   a. Strongly connected
   b. Weakly connected
   c. Connected
   d. Unconnected
11. How many nodes are in this network's largest weakly connected component?

4.11 Build a retweet network for a hashtag that represents a topic in which you are interested. Refer to the Chapter 4 Tutorial for usage of the Twitter API. Use the Twitter search API to retrieve recent tweets about the hashtag. Make sure you have `'result_type': 'recent'` in the search parameters. Extract at least 1000 tweets matching your search query; if there aren't that many tweets, start over and search for something else. Since this exceeds the maximum number of tweets that can be returned by the Twitter API with a single search request (100 at the time of writing), you will need to use pagination. Finally, build the retweet network from these tweets.

1. Draw the retweet network. To make it informative, follow these guidelines. (i) The nodes should be sized in proportion to their out-degree. (*Hint*: See Exercise 3.6 about how to get the degree sequence.) (ii) The links should be sized according to the number of tweets between the two users. (*Hint*: Use the `width` parameter of the `draw` function. The value of this parameter should be a list of edge weights, in the same order as given by the graph's `edges` method.) (iii) Singleton nodes and self-loops should be removed. (iv) Use your judgment whether to show only the largest connected component or all components.
2. What is the screen name of the most retweeted user? Also report the hashtag you used.
3. In a few sentences, report on some interesting observation about this retweet network.

4.12 Analyze the weighted network datasets available on the book's GitHub repository to study the relationship between degree and strength. For undirected networks, measure the Pearson correlation coefficient between degree and strength across all nodes. For directed networks, do the same for in/out-degree and in/out-strength. Do nodes with high degree also have large strength?

4.13   Consider a retweet network from one of the previous exercises, where link weights represent retweet counts. Prune the network by removing links with weights below a threshold ω.

1. Recall the definition of density for a directed network from Chapter 1. Draw a plot showing how the network density (on the y-axis) decreases as a function of the weight threshold ω (on the x-axis).

2. How much does the density decrease when you apply a threshold of $\omega = 3$ retweets?

3. What is the value of ω such that the density is reduced below half its initial value?

5 Network Models

mod·el: (*n.*) a simplified description, especially a mathematical one, of a system or process, to assist calculations and predictions.

We have seen that real networks of many different types share some common properties:

- They have short paths – it takes a few steps to go from any node to any other node.
- They have many triangles, reflected in high clustering coefficients.
- They have heterogeneous distributions of node and link variables, like degree and weights.

The next step in our investigation is to understand where such properties come from. How do nodes choose their neighbors? How are hubs generated? How are triangles formed? In this chapter we will answer these questions.

One way to study the origin of network characteristics is to formulate a *model* (i.e. a set of instructions used to assemble a network). The rules of a model incorporate intuitions, or hypotheses, about how network features emerge. By following the recipe of a model, we can build a network and compare it with real networks to see how they are similar or different. In this way, we can learn about the mechanisms that give rise to real-world networks.

Our exposition will trace the historical development of network science, presenting the classic models in order of their introduction. We discuss each model's failure to reproduce features observed in real networks, leading to the development of a new class of more realistic models. We will present simple mechanisms that allow us to generate model graphs having many basic features of real networks.

5.1 Random Networks

Suppose you have a bunch of disconnected nodes and wish to introduce some links. There are many ways to place the links between pairs of nodes. An "egalitarian" approach is to place them between randomly selected pairs of nodes. A network built in this fashion is called a *random* or *Erdős–Rényi* network (Box 5.1). For simplicity, let us formulate the model in an equivalent version proposed by Gilbert. The Gilbert model has two parameters: the number of nodes N and a *link probability p*,

Paul Erdős

The *Erdős–Rényi* random graph model is named after two mathematicians, Paul Erdős and Alfréd Rényi, who laid the foundations of random graph theory with several groundbreaking papers published jointly between 1959 and 1968.

Paul Erdős, shown in Figure 5.1, was an interesting character. He did not have a home, but he was not homeless. He would visit colleagues and stay at their homes while they worked on some mathematical problem together. Colleagues were happy to host him, as these visits were very productive professionally, often leading to prestigious scholarly publications. Once a theorem was proved or a paper written, Erdős would move to a new challenge, a new collaborator, and a new home.

Erdős worked on many different kinds of problems in addition to graph theory, and collaborated with over 500 colleagues. This makes him a hub in the math collaboration network, a social network discussed in Chapter 2.

Fig. 5.1   The mathematician Paul Erdős in 1992. Image adapted from commons.wikimedia.org/wiki/File:Erdos_budapest_fall_1992.jpg by Kmhkmh, used under CC BY 3.0 license.

describing how likely it is that a link is formed between any randomly chosen pair of nodes.[1]

The parameters of a random network model, as formulated by Gilbert, are the number of nodes N and the link probability p. The network can be constructed via the following procedure:

1. Select a pair of nodes, say i and j.
2. Generate a random number r between 0 and 1. If $r < p$, then add a link between i and j.
3. Repeat (1) and (2) for all pairs of nodes.

---

[1] The link probability is not to be confused with the rewiring probability that will be introduced in Section 5.2, although we use the letter p for both.

The main difference between the two formulations is that in the version by Erdős and Rényi the number of links of the network is fixed, whereas in the model by Gilbert it is variable. If we generate multiple networks following the procedure described in the above box, all using the same values for the number of nodes and the link probability, in general they will have different numbers of links, fluctuating around the average. However, when the number of nodes is sufficiently large, the fluctuations in the number of links are small.

To see what random networks look like for different values of the link probability, imagine a very large set of nodes, without links. Naturally, the system is totally fragmented into *singletons* (i.e. isolated nodes). Now let's add links at random, one at a time. What is going to happen? Clearly, an increasing number of pairs of nodes will get connected, and through them connected subnetworks will be formed. At some point the network will become connected, so that it will be possible to go from any node to any other node by following links. Therefore there must be a transition from configurations in which all subnetworks are relatively small, to a configuration where at least one of the subnetworks contains almost all the nodes. It is natural to expect that the subnetworks will grow smoothly and that the transition will happen gradually. Instead, Erdős and Rényi discovered that this transition is abrupt, and takes place as a specific density of links is reached. A *giant component* forms when $\langle k \rangle = 1$, that is, when each node has one neighbor on average.

In Figure 5.2 we show some configurations of the Erdős–Rényi graph for different values of the average degree. The largest connected component is very small before the transition

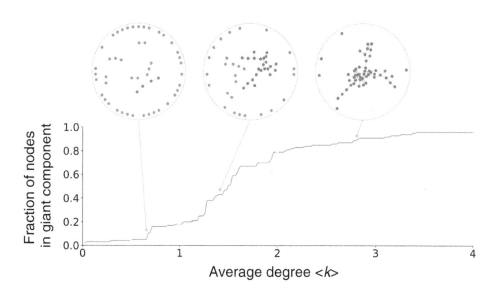

Fig. 5.2   Evolution of random networks for growing values of the average degree $\langle k \rangle$, which corresponds to the process of adding links to the system one at a time. The largest connected component, highlighted in red, is very small when the average degree is lower than one. Around $\langle k \rangle = 1$ a giant component grows very fast at the expense of the other, smaller components.

point and it grows rapidly with the average degree afterwards. The rest of the nodes are divided among small connected subnetworks. As the average degree gets even larger, the giant component "eats" all the remaining subnetworks and ends up including all nodes: the network becomes connected. Appendix B.2 presents a demonstration of the emergence of the giant component.

5.1.1 Density

Building a random network for a given value of the link probability is similar to the process of tossing a biased coin repeatedly and counting how many times we obtain heads or tails. The expected number of heads is proportional to the probability that the coin yields heads, and also proportional to the number of coin tosses. Analogously, the expected number of links in a random network is proportional to the link probability and to the number of node pairs.

Suppose a biased coin yields heads with probability p. For instance, if $p = 0.1$, we can expect that, on average, out of 10 tosses, we will obtain one heads and nine tails. If $p = 0.5$, we recover the familiar situation in which the coin is fair and we expect the same number of heads and tails. If $p = 0$, the coin never shows heads; if $p = 1$ instead, it never shows tails. The expected number of heads out of t tosses is then pt (i.e. a fraction p of the tosses). In our random network model, the number of "tosses" corresponds to the number of possible pairs out of N nodes, which is $\binom{N}{2} = N(N-1)/2$. Hence the expected number of links in a random graph is

$$\langle L \rangle = p \binom{N}{2} = \frac{pN(N-1)}{2}. \tag{5.1}$$

Remembering Eq. (1.6) that expresses the average degree of a network as twice the number of links divided by the number of nodes, we get the expected average degree $\langle k \rangle$ of a random network:

$$\langle k \rangle = \frac{2\langle L \rangle}{N} = p(N-1). \tag{5.2}$$

Equation (5.2) tells us that the expected average degree of a node in an Erdős–Rényi network is the fraction p of its $N - 1$ possible neighbors. Furthermore, plugging the expected number of links into the definition in Eq. (1.3), or plugging the expected average degree into Eq. (1.7), we find that the expected density of a random network is $\langle d \rangle = p$.

Intuitively, the link probability expresses the density of a random network: it is the expected ratio between the expected and maximum number of links. We know that real networks are usually sparse (i.e. they have very small average degree compared to the total number of nodes, and very small density). We conclude that for a random graph to be a good model of real networks, the link probability should be close to zero.

5.1.2 Degree Distribution

Suppose you have constructed a random network. What is its degree distribution? What we would like to know is the probability that a node has k neighbors. Since none of the nodes plays any special role in this model, we can just consider any one node, say i, and ask what is the probability that i has no neighbors, one neighbor, two neighbors, etc. Each of the remaining $N - 1$ nodes of the network could be a neighbor of i. By design, the decision on whether or not a link is placed between i and each other node is independent of the existence (or absence) of other links elsewhere. Each pair involving i has probability p of being connected, regardless of the rest of the network.

We are back to a coin-tossing problem, with a biased coin and a total number $N - 1$ of tosses. Our question turns into asking what the probability is that we obtain k heads out of $N - 1$ tosses if the probability of obtaining heads in each toss is p. This is given by the *binomial distribution*:

$$P(k) = \binom{N-1}{k} p^k (1 - p)^{N-1-k}. \tag{5.3}$$

In the limit of large N and constant (not too small) $pN \approx \langle k \rangle$, as in many real-world sparse networks, the binomial distribution is well approximated by a bell-shaped distribution with mean $\langle k \rangle$ and variance $\langle k \rangle$: the average degree is a good statistical descriptor of the distribution.

The resulting probability distribution for the degree in a random network is a bell-shaped curve with a prominent peak concentrated around the average degree $\langle k \rangle$, and rapidly decaying on both sides of the peak [Figure 5.3(a)]. The degree of most nodes is close to the average degree, and large deviations from it are very unlikely.

In Chapter 3 we have seen that the degree distribution of many real networks is rather different from such a distribution (Section 3.2), due to the presence of hubs (i.e. nodes with much larger degree than the average). In Figure 5.3(b) we plot the degree distribution of our world flight network. The heavy tail of the distribution covers over two orders of magnitude in degree; while many nodes have but a handful of neighbors, some hubs have hundreds. In Figure 5.3(c) we plot the distribution in double-logarithmic scale, and we compare it with that of Figure 5.3(a), which corresponds to a random network with the same number of nodes and links. Clearly, the random network model does not provide a good description of the distribution: the nodes have approximately the same degree, so there are no hubs. Such discrepancy is one of the reasons why we need more sophisticated network models.

5.1.3 Short Paths

Let us check whether random networks have short paths. We can use a simple argument to explore this question. In the previous section we have seen that the nodes have approximately the same degree. Let us suppose that they all have degree 10. If we start from any

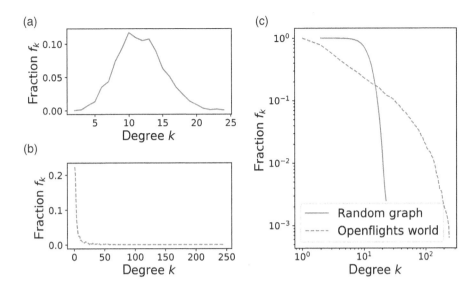

Fig. 5.3 Probability distribution of degree in a random network. (a) Degree distribution of an Erdős–Rényi random graph with the same number of nodes and links as the world flight network in our data collection: $N = 3179, L = 18,617$. (b) Degree distribution for the world flight network. (c) Comparison between the two distributions in (a) and (b) on a double-logarithmic scale.

node, there are 10 nodes attached to it. Each of them will also have 10 neighbors, and so on. So the number of reachable nodes grows exponentially with the number of steps: in two steps we can reach 100 nodes, in three steps 1000, and in a few steps we can reach every node in the network.

Assume the network is connected and all nodes have degree k. Within $\ell = 1$ step we reach k nodes. Each of these has $k - 1$ new neighbors if we exclude the root node we started from. So within $\ell = 2$ steps we can reach as many as $k(k - 1)$ nodes. Each of the new neighbors has in turn up to $k - 1$ new neighbors, so within distance $\ell = 3$ from the root we find $k(k-1)^2$ nodes, and so on. We conclude that, at distance ℓ from the root, we find up to $k(k-1)^{\ell-1}$ nodes. If k is not too small, we can approximate $k - 1 \approx k$ and the total number of nodes reachable within at most ℓ steps from any node is approximately k^ℓ. (This is actually an overestimate, because in reality neighbors of different nodes will occasionally coincide, whereas we are assuming that this never happens.) How far from a node do we have to go to reach all other nodes? The diameter ℓ_{max} such that the number of nodes reachable within at most ℓ_{max} steps from any node matches the total number of nodes N is given by

$$k^{\ell_{max}} = N, \tag{5.4}$$

from which we obtain

$$\ell_{max} = \log_k N = \frac{\log N}{\log k}. \tag{5.5}$$

It turns out that this is a good approximation of the diameter of the network, even when we consider neighborhood overlap and fluctuations in degree around $\langle k \rangle$. The slow logarithmic growth of ℓ_{max} with N indicates that distances within the network are small, even when the network size is very large.

The fact that the maximum distance to go from any node to any other node (the diameter) in a random network is small compared to the size of the network means that Erdős–Rényi networks indeed have short paths. To give a sense of how fast the number of reachable nodes grows with the distance from any node, let us consider the world's network of social contacts and imagine that it is a random network. If we take $k = 150$, which is the average number of regular contacts that humans can maintain (*Dunbar's number*), at distance five the number of reachable people is $150^5 \approx 75$ billion, a factor of 10 larger than the world population. So, in principle we could reach any individual in five steps or less, which is compatible with the result of Milgram's small-world experiment (Chapter 2).

5.1.4 Clustering Coefficient

As you will recall from Chapter 2, the clustering coefficient of a node measures the fraction of the node's pairs of neighbors that are connected to each other. The presence of a link between two neighbors closes a triangle with the focal node, so the clustering coefficient can also be interpreted as the fraction of triangles centered at the focal node, or the probability of closing a triangle.

On a random network, the probability that a pair of neighbors of a node is connected is p, as the link probability is the same for every pair of nodes, regardless of their having common neighbors or not. Naturally, the clustering coefficients of individual nodes may deviate a bit from p, but the average value across all nodes is well approximated by p. We have observed in Section 5.1.1 that p is a very small number if we are to describe real, sparse networks via the Erdős–Rényi model. It follows that the average clustering coefficient of these networks is very small — the model creates triangles with extremely small probability. In contrast, we know that real social networks have high clustering (Section 2.8). Therefore random networks are either unrealistically dense, or have unrealistically few triangles. We conclude that, if we wish to account for the remarkably high proportion of triangles observed in many real networks, we need a model with some specific rule to create triangles. Such models will be presented in Sections 5.2 and 5.5.3.

NetworkX has functions to generate random graphs according to the Erdős–Rényi and Gilbert models:

```
G = nx.gnm_random_graph(N,L) # Erdos-Renyi random graph
G = nx.gnp_random_graph(N,p) # Gilbert random graph
```

5.2 Small Worlds

As we have seen in the previous section, real networks are different from random ones. Erdős–Rényi networks do have short paths, but triangles are rare, resulting in average clustering coefficient values that may be orders of magnitude smaller than those measured in real networks.

In the late 1990s, Duncan J. Watts and Steven H. Strogatz introduced the *small-world model*, also known as the *Watts–Strogatz model*, which generates networks with both features — short paths and high clustering. Their idea is to start from a grid-like network where all nodes have the same number of neighbors, like the hexagonal lattice in Figure 5.4(a). Such a network has a high average clustering coefficient, as any pair of consecutive neighbors of each node are connected, forming a triangle with the node.

The internal nodes have degree $k = 6$ and clustering coefficient $C = 6/\binom{6}{2} = 6/15 = 2/5$. Border nodes have smaller degree $k = 4, 3, 2$ and even higher clustering coefficients, respectively $C = 3/\binom{4}{2} = 1/2$, $C = 2/\binom{3}{2} = 2/3$, and $C = 1/\binom{2}{1} = 1$. Therefore the average clustering coefficient is at least 2/5 and converges to 2/5 in the limit of infinite lattice ($N \longrightarrow \infty$).

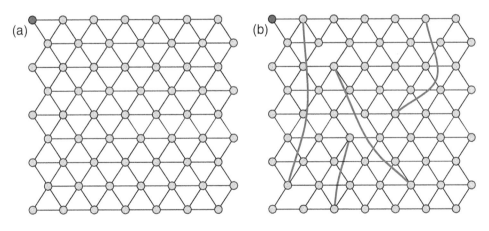

Fig. 5.4 Small-world networks. (a) A hexagonal lattice, a graph where each node has six neighbors (except on the boundary). There are many triangles, so nodes have a high clustering coefficient. Paths from one corner to another have to traverse many links, therefore the average shortest-path length is large. (b) Four links have been rewired to randomly selected nodes, which are typically further away from the original endpoints. These links (in red) are shortcuts, and allow us to reach remote parts of the network via a low number of hops. For instance, the shortest path from the blue node to the green node goes from 10 steps on the lattice down to six steps through one of the shortcuts. Since only a few triangles are disrupted by the rewiring procedure, the clustering coefficient remains high.

In contrast, the network has a high average shortest-path length. For example, nodes on opposite sides of the lattice can only be reached via paths traversing many links. The distances between nodes, however, could be reduced considerably by creating a few *short-cuts* — links joining portions of the network originally far from each other, like the red links in Figure 5.4(b). This can be done by selecting some of the initial links at random, preserving one of their endpoints and replacing the other endpoint with a node selected at random among all other nodes. Formally, the rewiring procedure applies to each link of the network with a *rewiring probability p*.[2] The number of rewired links is proportional to the rewiring probability.

The expected number of rewired links is pL, where p is the rewiring probability and L is the total number of links in the network. The special case $p = 0$ corresponds to the initial lattice, and the special case $p = 1$ yields a random network.

If the rewiring rate is very small (close to zero), little happens. If it is very large (close to one), the network becomes a random network á la Erdős and Rényi, as basically all links are rewired to random nodes, which is equivalent to placing links between randomly chosen pairs of nodes. In this scenario most triangles are destroyed and the clustering coefficient becomes very small. But if p is chosen to be neither too small nor too large, it is possible to achieve a trade-off, where there are enough shortcuts to make the distances considerably smaller on average, but not so many as to disrupt most of the triangles. In this regime, the paths are as short as those in random networks, while the average clustering coefficient decreases only slightly with respect to the initial lattice configuration, so that it can be comparable with that of real social networks.

In Figure 5.4 we start from a hexagonal lattice, but any network with high clustering coefficient can be used as an initial configuration. In their pioneering paper, Watts and Strogatz (1998) imagined a ring in which each node is connected to its k nearest neighbors. They considered the case $k = 4$, as shown in Figure 5.5(a). In this case the initial clustering coefficient is $C = 1/2$, because each node's neighbors form three triangles out of six possible. This is very high. One can rewire a link so that one of its attached nodes keeps the link while the other end of the link is attached to a randomly chosen node; this is the formulation in the original model. Alternatively, one can replace a link by connecting two random nodes, irrespective of their degree. Yet another variation is that random links can simply be added to the network, instead of rewiring the existing ones.

Figure 5.5(b) plots the average shortest-path length $\langle \ell \rangle_p$ and clustering coefficient C_p as a function of the rewiring probability p. There is a range of rewiring probability values between $p \approx 0.01$ and $p \approx 0.1$ (highlighted in Figure 5.5(b)) in which $\langle \ell \rangle_p \approx \langle \ell \rangle_1$ and $C_p \approx C_0$. In other words, the average shortest-path length from the model is close to that of an equivalent random network, and much lower than that of the lattice. At the same time,

---

[2] Note that the rewiring probability in the small-world model is not the same as the link probability in the random network model, despite the fact that we use the letter p for both. While this might be a bit confusing, it is a convention of the network science community that we follow; please be careful to interpret the parameter p based on the context of the model that is being discussed.

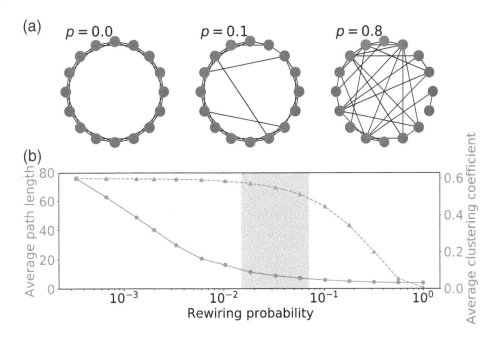

Fig. 5.5 Small-world model. (a) In the standard configurations of Watts–Strogatz model networks, we start from a ring lattice (left), where each node is connected to its four nearest neighbors, and progressively add shortcuts by rewiring links. (b) Decrease of the average shortest-path length and clustering coefficient as a function of the rewiring probability p. The extreme $p = 0$ is a lattice network as in the leftmost graph above, but with $N = 1000$ nodes. The extreme $p = 1$ is a random network with the same number of nodes and links. The shaded area highlights the values of p such that the average path length is almost as short as that of a random network, while the clustering coefficient is still almost as large as that of the lattice.

the clustering coefficient from the model is still close to that of the lattice, and much larger than that of the random network. Therefore, the Watts–Strogatz model is indeed capable of generating — with a suitable amount of randomness — networks endowed with two desired features: short paths and high clustering. A demonstration of this is presented in Appendix B.3.

The model cannot generate hubs, however. The degree distribution transitions from that of the initial lattice, where all nodes have identical degree, to that of a random network with the same number of nodes and links, where the degrees are concentrated around a characteristic value [Figure 5.3(a)]. Hence, for any value of the rewiring probability p, all nodes have similar degree; none accumulates a disproportional fraction of links. We need some other model ingredient to account for the emergence of hubs.

NetworkX has a function to generate graphs according to the small-world model of Watts and Strogatz:

```
G = nx.watts_strogatz_graph(N,k,p) # small-world model network
```

The degree sequence of a network is the list of degrees of its nodes, in the order of their labels. The degree sequence is a list of N numbers $(k_0, k_2, k_3, \ldots, k_{N-1})$, where k_i is the degree of node i. Note that the degree sequence determines the degree distribution, but the reverse is not true. Every permutation of a degree sequence leads to the exact same distribution, as for the distribution it does not matter which node has which degree, only how many nodes have a given degree.

5.3 Configuration Model

Let us focus on networks with realistic degree distributions. In Section 5.4 we will explore the mechanisms responsible for the existence of hubs. But first, let us answer the following question: given some degree distribution, can we construct a network whose nodes have exactly that degree distribution?

A simple solution is provided by the *configuration model*. This model actually pursues a more ambitious goal: generating a network whose nodes have an arbitrary *degree sequence*, with node 1 having degree k_1, node 2 degree k_2, and so on (Box 5.2). A degree sequence could be produced from a particular distribution we are interested in reproducing, or it could be taken from the nodes of a real network. Once we reproduce the sequence of all node degrees, we must have reproduced the corresponding degree distribution as well. In contrast, many degree sequences correspond to the same distribution. For example, two networks with distinct degree sequences (1,2,1) and (1,1,2) have the same degree distribution.

Suppose we have a set of nodes and their degree sequence. The first step is to assign to each node a number of *stubs* corresponding to the degree of the node, as in Figure 5.6(a). A stub is just a dangling link having the node as one of its endpoints, but not yet connected to a neighbor. The network is then constructed by the following iterative steps:

1. A pair of stubs is selected at random.
2. The chosen stubs are joined to each other, forming a link between the nodes attached to the stubs.

This routine is repeated until all stubs are joined in pairs. Naturally, for this to happen, there needs to be an even number of stubs (i.e. the sum of the degrees in the target sequence must be even). We see why this procedure achieves our aim: if a node has k stubs attached to it, it will eventually have k neighbors. Since the number of stubs attached to each node equals its degree, each node ends up having the desired degree. As illustrated in Figure 5.6(b–d), multiple networks can be created this way, depending on the sequence of pairs of stubs that are combined. Some outcomes may not be desirable if they violate constraints. For example, one may wish to exclude networks with multiple links between two nodes [Figure 5.6(c)] or with self-loops [Figure 5.6(d)].

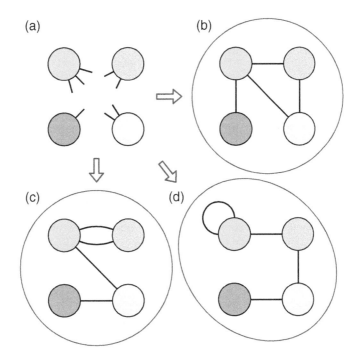

Fig. 5.6 Configuration model. (a) We start with nodes and stubs corresponding to a given degree sequence. (b–d) We can connect the stubs in different ways, leading to different networks with the given degree sequence.

The way links are formed is random by construction. So, the configuration model generates random networks with a prescribed degree sequence. This turns out to be very useful in the analysis of networks. In Chapter 3 we have seen that broad degree distributions are responsible for a number of peculiar properties and effects. However, there are network features that do not depend on the degree distribution alone.

Given a network, we might question whether a specific property is explained by the degree distribution alone. Through the configuration model we can generate randomized, or *shuffled*, versions of the network having the same degree sequence. Each of these configurations is a *degree-preserving randomization* of the original network, in that it preserves the degree sequence, but everything else is otherwise totally random. Now we can check whether the feature of interest is present in the shuffled network configurations. If it is there, then the feature must result from the degree distribution alone; otherwise there must be other factors behind it.

For instance, suppose that the feature we would like to investigate is the average clustering coefficient: can the clustered structure of a real social network be explained by its degree distribution? What we need to do is compute the clustering coefficient of a sufficient number of random configurations, derive the average and standard error, and check whether the value of the measure in the original graph is compatible with the estimate from the shuffled networks, within the error. If it is, we deduce that the triangles present in the network exist simply because of degree constraints. If it is much larger than the random estimate, as usually happens, the linking patterns of the network cannot be random; they must result from some mechanism that favors the formation of triangles.

NetworkX has a function for generating a network with a prescribed degree sequence via the configuration model:

```
G = nx.configuration_model(D)  # network with degree sequence D
```

Box 5.3   **Exponential Random Graphs**

It is of interest to study randomly generated networks sharing some common quantitative features, while differing in their detailed structure. On the one hand, they represent potential alternatives to the specific network configurations we encounter in the real world. On the other hand, they allow us to investigate the interplay between different structural properties. For instance, we might ask which values of the average clustering coefficient are compatible with a specific value of the density.

Exponential random graphs are classes of random networks subject to constraints. We define a class of networks based on a set of M network measures, x_m, $m = 1, \ldots, M$. We impose a constraint for each measure x_m: the average over all networks of the class must equal a specific value, say $\langle x_m \rangle = x_m^*$. Exponential random graphs are networks that satisfy these constraints while maximizing randomness. As it turns out, this allows us to define the probability $P(G)$ of selecting a network G of the class having measure values $x_1(G), x_2(G), \ldots, x_M(G)$:

$$P(G) = \frac{e^{H(G)}}{Z}, \tag{5.6}$$

with

$$H(G) = \sum_{m=1}^{M} \beta_m x_m(G), \tag{5.7}$$

where β_m is a parameter associated with measure x_m. The function Z ensures that $P(G)$ is a probability, so that $\sum_G P(G) = 1$.

Through the probabilities in Eq. (5.6) one can calculate the average of any network measure. In particular, we can express every constraint by setting the average of x_m to its desired value:

$$\langle x_m \rangle = \sum_G P(G) x_m(G) = x_m^*, \tag{5.8}$$

where the sum runs over all networks of the class. This yields a set of M equations with M variables, the parameters β_m. Solving these equations yields the values of the parameters. This specifies the model, which can then be used to calculate the average of any variable of interest. Although the constraints are imposed on averages, it turns out that for most exponential random graphs in a class, the value of any measure is close to its average.

The random network model, in the formulation by Gilbert (Section 5.1), is a special version of an exponential random graph with the single constraint that the networks must have a given average number of links.

The configuration model generates all possible networks having a given degree sequence, but we might as well impose other constraints. For instance, we could be interested in exploring all networks having a given number of triangles. The idea of generating networks with specific characteristics has led to the development of a broad class of network models called *exponential random graphs* (Box 5.3).

5.4 Preferential Attachment

The models explored so far are *static*. By that we mean that all the nodes of the network are there from the beginning; all we do is add (or rewire) links between them. Real networks, instead, are usually *dynamic*. Nodes and links appear and disappear. If we consider popular networks such as the Internet, the Web, Facebook, and Twitter, we notice that their size has been growing. Nodes may also disappear (e.g. old routers of the Internet or old Web pages). But the introduction of new nodes is more likely. This is the reason why realistic dynamic models typically incorporate some form of *network growth*. The dynamic procedure starts from an initial configuration, usually a very small clique of nodes. Then nodes are added one by one. Each new node is attached to a number of old nodes based on some rule, which is characteristic of the model (Figure 5.7).

Another limitation of the network models considered until this point is that they cannot explain the existence of hubs. To be more precise, the configuration model can generate hubs, but only by deciding *a priori* the degree of the nodes — that is not helpful in explaining how hubs emerge in the real world. The random network and small-world models do not give rise to hubs. The main reason is that in both cases the linking rules are basically egalitarian — the nodes choose their neighbors totally at random. This way, it is extremely unlikely for any one node to have an advantage over the other nodes and end up with many more neighbors than the rest. If we want to recover the hubs, it is thus necessary to introduce a mechanism that favors some nodes over others. Such a mechanism is called *preferential attachment*: the higher the degree of a node, the more links it will receive.

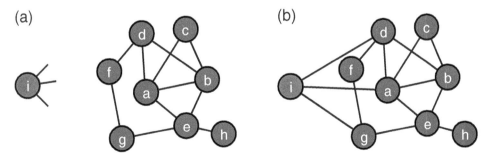

Network growth. The construction of a network is usually dynamic, with new nodes being added and connected to older nodes. (a) A new node **i** with three stubs is added to the system. (b) Each stub is attached to an older node, according to some rule, and the new node is incorporated in the network.

The underlying idea is simple. Suppose that you create a new Web page and wish to include some links to other pages. Our knowledge is necessarily limited to a tiny fraction of the trillions of pages on the Web. Most pages of which we are aware are likely to be popular, and as such they are linked from many other pages. Incidentally, this could be the reason why we discovered them: if a page has incoming links from many documents, it is more likely to be found via Web surfing or a search engine. Therefore, our choice of documents to link from our new Web page will favor popular and highly linked pages. Likewise, when we write a scientific article and we compile its bibliography, it is common to refer to papers that are frequently cited by other authors, as we are likely to bump into them when we read other articles and scroll through their references.

In the language of networks, a popular node is one with high degree, indicating that it has many neighbors. Preferential attachment means that high-degree nodes have a high probability of receiving new links. Such a criterion has been known in various contexts and under multiple names (Box 5.4). The best-known network growth model incorporating preferential attachment was proposed by Barabási and Albert in 1999, and is known as the *Barabási–Albert model*, or *BA model*, or *preferential attachment model*. It is a simple combination of growth and preferential attachment. At each step, a new node is added and connected to some existing nodes. The probability that a new node is attached to an old node is *proportional* to the degree of the old node. In this way, a node with degree 100 is 100 times more likely to receive a new link than a node with degree one. In Figure 5.7, for example, the chances that node **a** receives a link from node **i** would be double the chances of node **c** under preferential attachment.

Box 5.4   **Preferential Attachment**

The principle behind preferential attachment is simple: the more you have, the more you will receive! It is ancient, too. The first known reference can be found in the Gospel of Matthew (25:29): "*For to every one who has will more be given, and he will have abundance; but from him who has not, even what he has will be taken away.*" In this quote, the principle is summarized in the first sentence, the other one is its symmetric counterpart, stating that the less one has, the less one will have in the future. So *the rich get richer* and *the poor get poorer*. Preferential attachment is therefore also called the *Matthew effect*. Another common name is *cumulative advantage*.

The first scientific implementation of this principle was Pólya's *urn model*, which works as follows. An urn contains X white and Y black balls; one ball is drawn randomly from the urn and put back in, along with another ball of the same color. If X is much larger than Y, it is more likely that we extract a white ball than a black ball. If indeed we pick a white ball, at the end of the round there will be $X + 1$ white balls and Y black ones, which will give an extra advantage to white in the next round. Hence, the number of white balls will increase faster than the number of black balls.

Preferential attachment has been used to explain heavy-tailed degree distributions of many different quantities, like the number of species per genus of flowering plants, the number of (distinct) words in a text, the populations of cities, individual wealth, scientific production, citation statistics, and firm size, among others. Preferential attachment models were introduced by George U. Yule, Herbert A. Simon, Robert K. Merton, Derek de Solla Price, Albert-László Barabási, and Réka Albert.

We start from a complete graph with m_0 nodes. Each iteration of the algorithm consists of two steps:

1. A new node i is added to the network, with $m \leq m_0$ new links attached to it. The parameter m is thus the average degree of the network.
2. Each new link is wired to an old node j with probability

$$\Pi(i \leftrightarrow j) = \frac{k_j}{\sum_l k_l}. \tag{5.9}$$

The denominator in Eq. (5.9) is the sum of the degrees of all nodes (except i), and guarantees that the sum of all probabilities equals one, as it must be.

The procedure is repeated until the network reaches the desired number of nodes N. Box 5.5 shows how to select a node with the desired probability in Python.

Box 5.5   **Random Selection with a Probability Distribution**

It is often necessary to randomly choose nodes with probability proportional to some quantity. For example, in the case of a random network, we select a node to attach a link with uniform probability, meaning that each node has the same chance of being selected. In Python we can do this using the `random` module:

```
nodes = [1, 2, 3, 4]
selected_node = random.choice(nodes)
```

In other cases we need to select nodes with different probabilities. For example, in the case of preferential attachment (Section 5.4), at each step we need to choose a node with probability proportional to its degree. Or in the case of the fitness model (Section 5.5.2), selection is weighted by some more complicated function of degree and fitness. Fortunately, dealing with these cases is also easy as of Python 3.6. All we have to do is provide a second argument with a list of weights associated with the nodes. Suppose we wish to select a node according to its degree, as in preferential attachment. We can use the degrees as weights:

```
nodes   = [1, 2, 3, 4]
degrees = [3, 1, 2, 2]
selected_node = random.choices(nodes, degrees)
```

Node 1 ($k = 3$) is three times more likely to be selected than node 2 ($k = 1$). The `random.choices()` function allows one to randomly select from a population based on any given set of weights. The weights can be probabilities from a distribution, but it is not necessary — they do not need to sum to one. And they do not have to be integers. Do take care to ensure that the population and the weight sequences are aligned: the ith element in the population must correspond to the ith element in the sequence of weights.

At the beginning, all nodes have equal degree by construction. While new nodes and links are added to the system, the degrees of the nodes grow. However, the oldest nodes are there from the beginning, so they can receive links at any time, in contrast to nodes that enter the game much later. So the degrees of the oldest nodes exceed those of newer nodes, and this makes the former even more likely to attract new links in the future, at the expense of the latter, because of preferential attachment. Such *rich-gets-richer* dynamics generate the desired heterogeneity in the degree distribution, with the oldest nodes becoming network hubs. In Figure 5.8(a,c) we show a network constructed with the BA model, along with its degree distribution. We observe a heavy-tailed distribution, which confirms the existence of hubs. Appendix B.4 presents a demonstration of the model.

At this point you might wonder whether, for hubs to emerge, it is sufficient to have growth, without preferential attachment. After all, the initial nodes will have more time to collect links, whatever the linking criterion. For instance, suppose that each new node can pick as neighbor any randomly chosen node, regardless of its degree. Like before, we expect that the older the nodes, the larger their degrees. This is true, but we can see in Figure 5.8(b,c) that in this case the values of the degrees are not very different from each other, and the corresponding distribution does not have a heavy tail. We conclude that the combination of growth and random node selection does not do the job; preferential

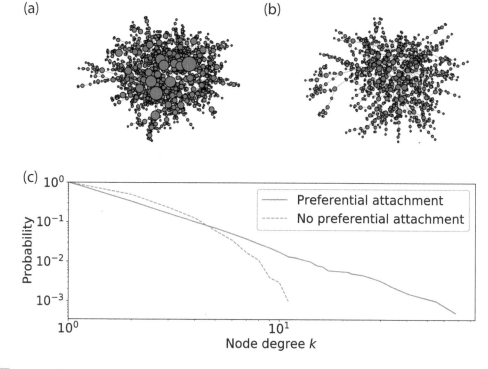

Preferential attachment. (a) Network generated with the BA model. It has $N = 2000$ nodes and average degree $\langle k \rangle = 2$. The size of a node is proportional to its degree; the large nodes are the hubs. (b) Network generated via a similar growth model, but with random rather than preferential attachment. There are no hubs. (c) Cumulative degree distribution of the networks in (a) and (b). The BA model generates a broad distribution, while the absence of preferential attachment leads to a narrower distribution, without hubs.

attachment is needed. Indeed, empirical studies have confirmed that preferential attachment is at work in the growth of many real networks.

NetworkX has a function to generate graphs according to the BA model:

```
G = nx.barabasi_albert_graph(N,m) # BA model network
```

5.5 Other Preferential Models

The BA model uses *linear* preferential attachment, in that the link probability is strictly proportional to the degree of the target node. Suppose that we relax this rule, and let the probability vary as some power of the degree. We call this *non-linear preferential attachment*.

The extension of the BA model using non-linear preferential attachment is identical to the original BA model, except that Eq. (5.9) expressing the probability that an old node j receives a link from a new node i is given by

$$\Pi_\alpha(i \leftrightarrow j) = \frac{k_j^\alpha}{\sum_l k_l^\alpha},$$

(5.10)

where the exponent α is a parameter. For $\alpha = 1$ we recover the standard BA model. What happens when $\alpha \neq 1$? There are two different scenarios:

1. If $\alpha < 1$, the link probability does not grow with degree as fast as in the BA model, so the advantage of high-degree nodes over the others is not as big. As a result, the degree distribution does not have a heavy tail — the hubs disappear!
2. If $\alpha > 1$, high-degree nodes accumulate new links much faster than low-degree nodes. As a consequence, one of the nodes will end up being connected to a fraction of all other nodes. The effect is even more extreme when $\alpha > 2$, in which case we observe a *winner-takes-all* effect: a single node may be connected to all other nodes, which have approximately the same, low degree.

Depending on the value of the exponent that expresses the power of the degree, we either end up without hubs (sub-linear preferential attachment) or with one super-hub (super-linear preferential attachment). Either way, non-linear preferential attachment fails to generate hubs as observed in real-world networks; linear preferential attachment is the only way to go. This exposes a fundamental fragility of the BA model, as the necessity of strict proportionality between link probability and degree appears unrealistic. Luckily, as we will see in Section 5.5.4, there are natural linking mechanisms that implicitly induce linear preferential attachment.

In addition to the dependency on linear preferential attachment, the BA model has other limitations:

- It yields a fixed pattern for the degree distribution. The slope of the preferential attachment curve in Figure 5.8(c) is the same for any choice of the model parameters. Real degree distributions could decay faster or more slowly.
- The hubs are the oldest nodes; new nodes cannot overcome them in degree.
- It does not create many triangles. The average clustering coefficient is much lower than in many real networks.
- Nodes and links are only added; in real networks they can also be deleted.
- Since each node is attached to older nodes, the network consists of a single connected component. Many real networks have multiple components.

Next we present more sophisticated models of network growth that address some of these limitations.

5.5.1 Attractiveness Model

Preferential attachment has a subtle pitfall: what happens if a node has no neighbors? Its degree is zero, so the probability that it will get links is also zero: the node will never have neighbors! So, if we started from an initial core of nodes with no neighbors, the BA model would collapse, as the new nodes could not be attached to any of the old nodes. The standard initial configuration of the BA model consists of a complete graph, so every node has neighbors and this problem does not occur, but ideally a model should work with different choices of the initial condition. If we consider directed networks, and assume that the link probability depends only on the in-degree, the problem occurs irrespective of the initial configuration. Each new node has initially in-degree zero, as it comes with outgoing links and can only receive incoming links from newer nodes. Therefore no new nodes can receive incoming links.

Fortunately, there is a simple way out of this. Instead of having a link probability that is strictly proportional to the degree, we can modify the rule slightly. The idea, originally proposed by Derek de Solla Price in the context of citation networks, is that a node receives links because of its degree, but also because it has an intrinsic attractiveness. In the *attractiveness model*, the link probability is *proportional to the sum of degree and a constant attractiveness*.

The attractiveness model is a slightly modified version of the original BA model, in which Eq. (5.9) expressing the probability that an old node j receives a link from a new node i is replaced by

$$\Pi(i \leftrightarrow j) = \frac{A + k_j}{\sum_l (A + k_l)}, \qquad (5.11)$$

where A is the attractiveness parameter and can take any positive value. The case $A = 0$ yields the BA model.

For any value of the attractiveness parameter A, the model builds networks with heavy-tailed degree distributions. The slope of the distribution depends on A. This way, the model is able to match degree distributions of multiple real networks, unlike the BA model.

5.5.2 Fitness Model

As we have seen in Section 5.4, in the BA model the hubs are also the oldest nodes. This feature is not realistic. In the Web, for example, there can be pages that are created long after others, but that end up being more popular and attracting more hyperlinks. Take Google, for example. It was created in 1998, when there were already millions of sites, but it ended up being the most popular Web hub. Likewise, in the scientific literature the most cited papers are not the oldest ones: occasionally new groundbreaking papers surpass many earlier publications.

This happens because nodes (websites, papers, social media users, and so on) have their own individual appeal that may boost the rate at which they accrue links, giving them an edge over much older nodes. Such appeal is only partly, and indirectly, reflected by their degree. The attractiveness parameter of the model described in the previous section is the same for all nodes, so it does not allow us to discriminate among nodes and to introduce discrepancies in their rate of degree growth. Therefore, in the attractiveness model, as in the BA model, hubs are still the oldest nodes.

To allow for the possibility that new nodes become hubs, Bianconi and Barabási proposed a *fitness model* in which each node has its own individual appeal, called *fitness*. The fitness values are intrinsic features of the nodes; they do not change over time. The link probability is proportional to the *product* of the degree and the fitness of the target node.

> The fitness model is similar to the BA model, but each node i is assigned a fitness value $\eta_i > 0$ generated from some distribution $\rho(\eta)$. Then at every step, each new link from a new node i is wired to an old node j with probability
>
> $$\Pi(i \leftrightarrow j) = \frac{\eta_j k_j}{\sum_l \eta_l k_l}. \qquad (5.12)$$
>
> If all nodes have identical fitness, the model reduces to the BA model, as the constant η is a factor that cancels out between numerator and denominator of Eq. (5.12), returning the standard prescription of preferential attachment.
>
> If the fitness distribution $\rho(\eta)$ has *infinite support* (i.e. η can take arbitrarily large values), then there is a winner-takes-all effect with the highest-fitness node being linked to most modes. But if the fitness distribution $\rho(\eta)$ has *finite support* (i.e. η has a finite maximum value), then the degree distribution of the model has a heavy tail. An example of this case is the uniform distribution in the unit interval. In Python we can draw fitness values with uniform distribution using the `random()` function.

The fitness model generates networks with two desired properties. First, as long as the fitness values are bounded, the network has multiple hubs. Second, high fitness allows a node to compete with all its peers, regardless of their age and status. This is because the nodes increase their degree at a rate determined by their individual fitness. Therefore the nodes with the largest fitness values eventually achieve the largest degrees, regardless of when they are introduced in the system.

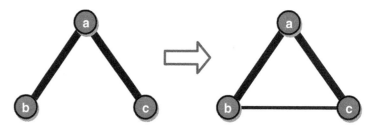

Fig. 5.9 Strong triadic closure. An individual **a** has strong connections, indicated by the thick links, with **b** and **c**. According to Granovetter's principle of strong triadic closure (Box 5.6), there must be or will eventually be at least a weak connection between **b** and **c**.

5.5.3 Random Walk Model

Networks built with the BA model have very low clustering coefficients. To see why, recall that in order to have many triangles, it is necessary that links join pairs of nodes with at least one common neighbor. For instance, if nodes **b** and **c** are both attached to **a**, a link between **b** and **c** will close the triangle **abc** (Figure 5.9). In the BA model, however, the probability of a node receiving a link is proportional to its degree, regardless of whether the new pair of neighbors have a common neighbor or not. This is why triangles are formed rarely. To increase the production of triangles it is necessary to introduce a mechanism that favors the creation of links between nodes with common neighbors.

The formation of a triangle via the addition of a link is called *triadic closure* and is perhaps the main mechanism explaining the formation of links in social networks (Box 5.6). That should not be surprising: many people we know have been introduced to us by common acquaintances. The mechanism has several model implementations. Here we discuss a very intuitive one called the *random walk model*. The idea is that in addition to creating random connections, we also connect to a new neighbor's neighbors — in a social network, to a new friend's friends.

The random walk model can start from any small network. Each iteration of the algorithm consists of the following steps:

1. A new node i is added to the network, with $m > 1$ new links attached to it.
2. The first link is wired to an old node j, chosen at random.
3. Each other link is attached to a randomly selected neighbor of j, with probability p, or to another randomly selected node, with probability $1 - p$.

The parameter p is the probability of triadic closure, because by setting a link between i and a neighbor of j, say l, we close the triangle (i, j, l). If $p = 0$, there is no triadic closure and new nodes choose their neighbors entirely at random. When $p = 1$ all links except the first one are wired to neighbors of the initially selected old node, thus closing triangles.

Box 5.6 **Triadic Closure and the Strength of Weak Ties**

In 1973, sociologist Mark S. Granovetter, after a long struggle with editors, published a paper entitled "The strength of weak ties" that would become the most cited article in sociology. The paper set forth a tight relationship between three fundamental features of social networks: triangles, link weight, and communities.

Granovetter introduced the principle of *strong triadic closure* for how links are formed in social networks. If person **a** has strong (high-weight) connections with two individuals **b** and **c**, it is very likely that **b** and **c** are friends or that they will eventually become friends. There may be various reasons. If **b** and **c** spend a lot of time with **a**, it is likely that they will eventually meet through **a**. Also, since **a** is a good friend of both, **b** and **c** will be inclined to trust each other. Finally, if **b** and **c** keep ignoring each other it might be a source of stress for the group. Strong triadic closure prescribes that there must be a link between **b** and **c**, so **a**, **b**, and **c** would form a triangle (Figure 5.9). This formalizes the relationship between triangles and link weights.

A social community is a circle of people having a lot of interactions with each other, because of family ties, work relationships, and so on. (We discuss communities in Chapter 6.) Granovetter argued that links with large weights, signaling *strong ties* between individuals, are most likely to be found within the same community, while low-weight links (*weak ties*) tend to lie between communities. The intuition is that people in different circles have limited contact. Granovetter provided an argument to support his theory. Suppose that there is a strong link between two individuals **a** and **b** belonging to different communities. Each of them is likely to have strong connections with other members of their own community. Let us suppose that **a** is a close friend of **c**. Due to strong triadic closure, there must be a link between **b** and **c** as well, because **ab** and **ac** are strong ties. But a link connecting two communities can hardly be a side of triangles, because otherwise the communities are not well separated, so **ab** must be weak. On the contrary, since there are many strong ties between members of the same community, there will be many triangles within communities. This argument suggests an interplay between communities and link weights, as well as between communities and triangles. Despite their low weights, weak ties are critical for the structure of social networks, because they connect communities to each other, enabling the spread of information across the network.

The model is illustrated in Figure 5.10. It creates networks with a number of triangles that can be varied by tuning a parameter p.[3] The largest density of triangles is obtained when $p = 1$. In addition, if p is not too small, the model generates heavy-tailed degree distributions. This is due to the process of triadic closure. By choosing a neighbor of an old node, we are just selecting a link of the network. As you may recall from Section 3.3, if we select a link at random, the probability that an endpoint of that link has a given degree is proportional to the degree. So, old nodes will receive links with a probability that is proportional to their degrees, exactly as in preferential attachment.

The mechanism used in the random walk model is more intuitive than the one in the preferential attachment model, as it does not assume that new nodes have knowledge of the degrees of old nodes. The model simply explores the network in a random fashion, and

---

[3] Again, do not confuse this parameter with those in the random network and small-world models that use the same letter p.

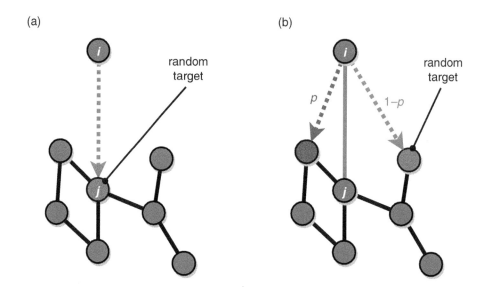

(a)   (b)

random target   random target

Fig. 5.10 Random walk model. (a) A new node i is attached to a randomly chosen node j. (b) Each additional link brought by i is attached to a neighbor of j with probability p, which leads to the formation of triangles. Otherwise the link is attached to a randomly chosen node.

the nodes are "discovered" with a frequency proportional to their degrees. In other words, the process of *triadic closure implicitly induces preferential attachment*. From this point of view it is basically equivalent to *link selection* (i.e. choosing a link at random and attaching the new node to one of the endpoints of the link). The difference is that in triadic closure the new node is attached to both endpoints of the selected link, but the resulting degree distributions are similar in both cases. Therefore, the strict proportionality between link probability and degree, necessary to generate broad degree distributions, can be enforced by simple mechanisms based on random choices.

Finally, when p is large enough that there is a sufficiently high density of triangles, the random walk model produces networks with community structure, which we will discuss in Chapter 6. The relationship between triangles and communities is well known in the network science literature (Box 5.6).

5.5.4 Copy Model

In social networks, triadic closure implies that an individual *copies* the contacts of somebody else. This kind of copying mechanism can take place in other contexts as well. Some examples:

- Gene duplication is a process through which new genetic material is generated during molecular evolution. Consider the protein–protein interaction network, where each node represents the protein expressed by a gene. When a gene is duplicated, the new gene/protein node will be interacting with the same proteins as the original node in the protein interaction network. So the links to those nodes are copied.

- Scholars often discover new articles in the lists of references of publications they read, and cite them in their own papers. In doing so, they copy (some of) the citations of other publications.
- While browsing online, Web content creators may discover relevant Web pages such as authoritative sources or hub pages that provide lists of resources. By linking to these pages from newly created pages, authors copy the hyperlinks leading to them.

These scenarios are captured by the *copy model*. It is similar to the random walk model discussed earlier: a new node gets wired either to a randomly selected old node, with some probability, or else to its neighbors. However, there is no triadic closure in the copy model; the new node does not get attached simultaneously to a node and (some of) its neighbors. Therefore we obtain networks with hubs, but with few triangles.

5.5.5 Rank Model

Preferential attachment implies that the nodes have a perception of how important other nodes are, as it requires knowledge of their degrees during network growth, to estimate the link probability and properly distribute the links. We may wonder whether we can operate without knowing the degrees of the nodes. In Sections 5.5.3 and 5.5.4 we have seen that triadic closure and link selection/copy are viable strategies. Here we consider a different approach.

In realistic settings, it is more common to have a perception of the *relative value* of things, rather than of their absolute value. We can state with great confidence that Bill Gates is much richer than any of the authors of this book, even though we ignore the exact amount of his wealth. The claim that we are able to rank the nodes of a network based on a specific variable (e.g. degree or age) is thus more believable than the claim that we can accurately estimate the values of the variable. This is the idea at the basis of the *rank model*.

We keep the nodes ranked by one of their properties, say the degree. Then we select nodes to receive new links with probability proportional to some inverse power of their rank. The top node will have the highest chance of receiving a link, followed by the node ranked second, third, and so on. How the link probability decays with the rank is determined by an exponent parameter.

The rank model can start from any small graph with m_0 nodes. A node property, such as the degree, age, or some measure of fitness is selected to rank the nodes. Each iteration of the algorithm consists of the following steps:

1. All nodes are ranked based on the property of interest. Nodes are assigned ranks $R = 1, 2$, etc. Node l receives rank R_l.
2. A new node i is added to the network, with $m \leq m_0$ new links attached to it.

3. Each new link from i is wired to an old node j with probability

$$\Pi(i \leftrightarrow j) = \frac{R_j^{-\alpha}}{\sum_l R_l^{-\alpha}}, \tag{5.13}$$

where the exponent $\alpha > 0$ is a parameter.

Nodes may have to be re-ranked at each iteration if the ranking property depends on the links from new nodes joining the network, as happens for example when nodes are ranked by degree.

A top-ranked node (small rank) is more likely to receive a new link than a poorly ranked one (large rank). If the variable used for the ranking is degree, this means that high-degree nodes have higher chances of attracting new links than low-degree nodes, as in the BA model. However, the actual link probability values are different because they depend on the ranks, not on the degrees.

As it turns out, the rank model generates networks with heavy-tailed degree distributions, for any property used to rank the nodes and any value of the exponent parameter. By tuning the exponent, it is possible to vary the shape of the distribution, and to reproduce the empirical distributions observed in many real-world networks.

Hubs are created even if nodes have partial information on the system, in that they are aware of the existence of only a fraction of the nodes. This reflects a familiar scenario. Imagine you are writing a Wikipedia article and wish to link to relevant news articles. You might use a search engine to identify pages to link. The search engine presents pages ranked by relevance to your query. You link to the top result with high probability; to the second result with half that probability; to the third with one-third probability; and so on. You might not even bother looking past the first page of results. Your article will be a new Wikipedia node, linking to old nodes according to a procedure closely resembling the rank model. This can help explain the emergence of popular hubs on the Web.

5.6 Summary

Network models help us understand the basic mechanisms that are responsible for the characteristic structural features observed in a real network. The basic ingredients of network models are the rules that determine how nodes get attached to each other. Below we list the lessons we have learned from the models reviewed in this chapter.

1. In random networks generated by the Erdős–Rényi model, every node has the same probability of becoming a neighbor of any other node. These networks have short paths, but there are very few triangles and no hubs.
2. The small-world model modifies an initial lattice structure with high average clustering coefficient by creating some random shortcuts between the nodes. A few shortcuts suffice to drastically reduce the distances between the nodes, inducing the small-world

property, while the clustering coefficient remains high. The model is unable to create hubs.

3. The configuration model generates networks with any predefined degree sequence. The structure is therefore imposed "by hand," it is not explained by the model. The configuration model is often used as a baseline, to check whether any property of a network is due to its degree distribution alone or to other factors. This can be done by comparing the property of interest in the original system and in randomized networks with the same degree sequence, created by the model.

4. Realistic network models include network growth, in that nodes and links are added to the graph over time. This matches the evolution of many real-world networks, like the Internet, the Web, etc.

5. Preferential attachment is the key mechanism to explain the emergence of hubs: the higher the degree of a node, the higher the probability that it will be connected to other nodes.

6. The Barabási–Albert model, with its combination of network growth and preferential attachment, yields networks with heavy-tailed degree distributions, therefore explaining the emergence of hubs.

7. Preferential attachment can be induced implicitly by simple processes involving random choices, like triadic closure and link selection.

8. Several models have been proposed to overcome the limitations of the Barabási–Albert model, by introducing ingredients such as attractiveness, fitness, triadic closure, and ranking.

5.7 Further Reading

The random graph model was introduced in the same year by Erdös and Rényi (1959) and Gilbert (1959), though the idea was put forward in an earlier paper by Solomonoff and Rapoport (1951). The Gilbert model is often erroneously attributed to Erdős and Rényi in the literature. We refer to the average number of regular contacts that humans can maintain as *Dunbar's number* (Dunbar, 1992).

The small-world model was developed by Watts and Strogatz (1998). Molloy and Reed (1995) proposed the configuration model. Exponential random graphs were introduced by Holland and Leinhardt (1981).

Barabási and Albert are usually credited with the preferential attachment model, often called the Barabási–Albert or BA model (Barabási and Albert, 1999). Other scholars had proposed the model before; the closest ancestor was a paper by Price (1976). Non-linear preferential attachment was investigated by Krapivsky *et al.* (2000) and Krapivsky and Redner (2001). Attractiveness was added to preferential attachment by Dorogovtsev *et al.* (2000). The fitness model was proposed by Bianconi and Barabási (2001).

Granovetter (1973) authored the pioneering paper "The strength of weak ties." The random walk model was introduced by Vázquez (2003). The copy model was an idea of Kleinberg *et al.* (1999) that emerged during early studies of the Web graph. Several authors proposed gene duplication models (Wagner, 1994; Bhan *et al.*, 2002; Solé *et al.*, 2002; Vázquez *et al.*, 2003a). The rank model was developed by Fortunato *et al.* (2006).

Exercises

5.1 Go through the Chapter 5 Tutorial on the book's GitHub repository.[4]

5.2 What is the difference between the random graph by Erdős and Rényi and that by Gilbert?

5.3 Suppose we want to construct a random graph with 1000 nodes and about 3000 links. Give a value of the link probability p that could lead to this outcome.

5.4 Suppose you are making a random network with 50 nodes, and you want the average node degree $\langle k \rangle$ to be 10. What approximate value for p would you use in this case?

5.5 Given a random network of 50 nodes and average node degree $\langle k \rangle = 10$, which of the following is likely to be closest to the average path length of that network?
   a. 1.5
   b. 2.0
   c. 2.25
   d. 2.5

5.6 Consider the family of random networks with average degree $\langle k \rangle = 10$. How many nodes are we likely to need in order to generate such a network with average path length $\langle \ell \rangle = 3.0$? (*Hint*: If you use a guess and check strategy, make sure that $\langle k \rangle$ stays the same for the various values of N. Each will involve a different value of p.)
   a. 60
   b. 100
   c. 250
   d. 500

5.7 Build a random network with 1000 nodes and $p = 0.002$. Plot its degree distribution. (*Hint*: We show how to plot a distribution in the Chapter 3 Tutorial.) Answer the following questions:
   1. What is the largest degree of the network?
   2. What is the *mode* of the distribution (i.e. the most common value of the degree)?
   3. Is the network connected? If not, how many nodes are in the giant component?
   4. What is the average clustering coefficient? Compare it with the link probability p
   5. What is the network diameter?

5.8 Consider the following process. First, start with a collection of N nodes and no links. This is clearly a disconnected network. Then, one by one, add a link between

---

[4] github.com/CambridgeUniversityPress/FirstCourseNetworkScience

two nodes not already connected to each other. Continue until you have a complete network. How many steps are in this process?

5.9 Consider the step-by-step process from the previous question. At each step in the process, suppose you record the size of the largest component. Which of the following tends to be true about the sequence of largest component sizes as you add edges?
   a. Increases slowly at the beginning of the sequence, increasing very fast at the end of the sequence
   b. Increases slowly until some threshold, increases very quickly for a short time, and then increases slowly afterwards
   c. Increases at a constant rate from the beginning to the end of the sequence
   d. Increases very fast at the beginning of the sequence, then tapers off and increases slowly until the end
   e. Increases and decreases randomly

5.10 Reproduce the plot of Figure 5.2 for networks with 1000 nodes. (*Hint*: Use the NetworkX function to generate random networks.) Use 25 equally spaced values of the link probability, in the interval $[0, 0.005]$. For each value generate 20 different networks, compute the relative size of the giant component and report the average and the standard deviation in the plot.

5.11 What is a reason why random networks are not good models of social networks?
   a. Random networks are typically not connected
   b. Random networks have small average shortest-path lengths
   c. Nodes in random networks have very different degrees
   d. Random networks have low clustering coefficients

5.12 Consider a ring-like lattice like the one in Figure 5.5(a), with 100 nodes each connected to its four nearest neighbors (two on either side of it). What is the average clustering coefficient? Does the network size matter? (*Hint:* Given the symmetry, it is sufficient to calculate the clustering coefficient of any node.)

5.13 The Watts–Strogatz model is useful in capturing which property of real-world social networks not found in Erdős–Rényi random graphs?
   a. Short average path lengths
   b. Long average path lengths
   c. Low clustering coefficient
   d. High clustering coefficient

5.14 Reproduce the plot of Figure 5.5(b), by calculating the average shortest path ($\langle \ell \rangle$) and the average clustering coefficient (C) for Watts–Strogatz networks constructed for different values of the rewiring probability p. Take 20 equally spaced values of p between 0 and 1. For each value of p, build 20 different networks and compute the average of $\langle \ell \rangle$ and C. To plot the two curves on a common y-axis, you can normalize the values by dividing them by the corresponding values for $p = 0$.

5.15 Build Watts–Strogatz networks with 1000 nodes, $k = 4$, and these values for the rewiring probability: $p = 0.0001, 0.001, 0.01, 0.1, 1$. Compute and compare their degree distributions, by plotting them in the same diagram.

5.16 Consider the US airport network (USAN). Create a randomized version of it (RUSAN) using the configuration model. To do that, take the degree sequence of the network and apply the `configuration_model()` function of NetworkX. Carry out the following tasks:
1. Verify that the degree distribution is identical to that of the USAN
2. Compare the average shortest paths of USAN and RUSAN. How do you interpret the difference in the values?
3. Compare the average clustering coefficients of USAN and RUSAN. How do you interpret the difference in the values?

5.17 Which of the following features would you expect to find in a BA but not in a random network with the same number of nodes and edges?
a. Nodes with degree greater than one
b. Hub nodes with degree many times larger than that of a typical node
c. Short average path lengths
d. Long average path lengths

5.18 Build a BA network with 1000 nodes and $m = 3$. Carry out the following tasks:
1. Plot the degree distribution of the network, in double-logarithmic scale
2. Derive the average degree, see how it compares with m, and interpret the result
3. Calculate the average clustering coefficient
4. Verify that the graph is connected
5. Calculate the average shortest path

5.19 Build an Erdős–Rényi random graph with the same number of nodes and links as the BA network in the previous exercise.
1. Derive the degree distribution and compare it with that of the BA network, by drawing them in the same plot, in double-logarithmic scale
2. Calculate the average clustering coefficient and average shortest path length and compare them with the corresponding values for the BA network. Interpret the results

5.20 The attractiveness and fitness models are based on the same idea that nodes have an intrinsic appeal, unrelated to their degree. What are the differences between the two models?

5.21 In the fitness model, suppose that the fitness of a node coincides with its degree. Could you guess what kind of degree distribution the resulting network will have? (*Hint*: The discussion on non-linear preferential attachment in Section 5.5 might help.)

5.22 Give a reason why networks generated by the BA model do not have many triangles.

5.23 If you use an online social network such as Facebook, Instagram, or LinkedIn, consider your links in this network: how many are strong ties and how many are weak ties?

5.24 Link selection consists of choosing a link and wiring a new node to one of its endpoints. Suppose that we wire the new node to both endpoints. What would be the difference from the random walk model? And how would networks generated by the two models differ?

5.25 Suppose that we wish to build a network such that there is a relevant amount of squares (cycles of length four). Based on what you learned about triadic closure, could you suggest a mechanism that incentivizes the formation of squares?

5.26 Consider two versions of the rank model, with different ranking criteria. In the first version nodes are ranked by age (the time elapsed since they were added to the network). In the second version nodes are ranked by their degree. Is there a difference between networks generated by the two models, and if so, what is it?

5.27 The `socfb-Northwestern25` network in the book's GitHub repository is a snapshot of Northwestern University's Facebook network. The nodes are anonymous users and the links are friend relationships. Load this network into a NetworkX graph; be sure to use the proper graph class for an undirected, unweighted network. Once you measure the number of nodes and links, use `nx.gnm_random_graph()` to create a separate random network with the same number of nodes and links as the Facebook graph. Use this random network to answer the following questions:

  1. What is the 95th percentile for degree in the random network (i.e. the value such that 95% of nodes have this degree or less)?
  2. We are dealing with a random network, so some properties are going to differ somewhat each time one is generated. True or false: Given fixed parameters N and L, all random networks created with `nx.gnm_random_graph()` will have the same mean degree.
  3. Which of the following shapes best describes the degree distribution in this random network?
     a. Uniform: node degrees are evenly distributed between the minimum and maximum
     b. Normal: most node degrees are near the mean, dropping off rapidly in both directions
     c. Right-tailed: most node degrees are relatively small compared to the range of degrees
     e. Left-tailed: most node degrees are relatively large compared to the range of degrees
  4. Estimate the average shortest-path length in this random network using a random sample of 1000 pairs of nodes.
  5. What is the average clustering coefficient of this random network? Answer to at least two decimal places.

## 6

Communities

clus·ter: (*n.*) a group of similar things or people positioned or occurring closely together.

When you look at the layout of a network, one of the first things you may notice is that nodes are grouped in *communities*, also called *clusters* or *modules* — sets of nodes with a relatively higher density of connections within than between them (Figure 6.1).

Communities often tell us how a network is organized and what functions it serves. For example, dense clusters of connected neurons in the brain are often synchronized in their firing patterns. In a protein–protein interaction network, groups of connected (interacting) proteins are typically associated with a particular biological function within the organism. Sometimes we can infer the role of an unknown gene with respect to a complex disease by looking at the cluster it belongs to in a gene regulatory network. On the Web, as we discussed in Section 4.2.5, clusters of pages with many hyperlinks pointing to each other usually identify a topic. And in social networks, as discussed in Section 2.1, communities of friends share important features, such as political beliefs. Social communities can have significant influence on public opinion. For example, when looking at the diffusion network for politically charged memes on Twitter (Figure 0.3), we immediately note that people are divided across two distinct communities that do not interact much with each other.

A similar picture is offered in Figure 6.2, showing the retweet network of political memes in the United States. Again we observe two mostly segregated communities. When we inspect some of the users, it becomes clear that the two clusters are aligned with political right and left. As discussed in Section 4.5, this scenario is sometimes called an *echo chamber* or *filter bubble* to indicate that a person is only exposed to people with similar ideas and beliefs. While it is normal for us to have friends like us in real life, with online social networks and social media it is easier to filter out different views, as we are encouraged to connect with people who are similar to us or with whom we already share common friends. We also have tools to easily mute or ignore people with whom we disagree, which is a little harder to do in real life. It has been theorized that when opinions are unchallenged, biases are reinforced and polarization may ensue.

Knowing the community structure of a network allows us to classify the nodes based on the position they have in their own clusters. Nodes that are fully embedded in a cluster, in that their neighbors belong to the same cluster, represent the core of the group, as they do not mix with members of other groups. Nodes lying at the boundary of a community have neighbors both inside and outside their group and serve as gatekeepers between different parts of the network. As such, they play an important role for diffusion processes taking

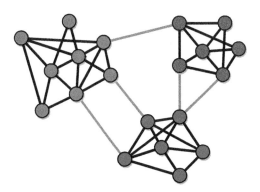

Fig. 6.1 A network with three communities, indicated by node colors.

place in the network. If we want to stop a rumor or fake news spreading in a social network, halt the spreading of an epidemic in a contact network, or make sure critical information reaches all communities, those are the nodes we need to control.

Given the importance of communities in understanding the functions of a network and of individual nodes, it is critical to be able to detect communities in a network. Sometimes communities are very evident and all we need to do to see them is lay out the nodes on a plane so that connected nodes are positioned close to each other. That is the idea behind the popular *force-directed network layout algorithms* described in Chapter 1. For example, in Figure 6.2, because people in a community (liberal or conservative) are densely connected to each other, they end up clustered together in the layout. Many networks of interest, however, are much larger than those for which it is possible to produce meaningful visualizations. And even in many small systems, visualization does not help to identify the clusters. So it is necessary to develop algorithms that can automatically discover the communities starting from the knowledge of the network structure and possibly other inputs (e.g. the desired number of clusters).

The problem of identifying communities in networks is a truly interdisciplinary topic, and as such it takes different names: *community detection*, *community discovery*, and *clustering*,[1] among others. Grouping nodes into communities is considered an *unsupervised* classification task, in that we do not have precise prior knowledge or examples of what the resulting partitions should look like. Indeed, there is no unique definition of community. The natural intuition is that there should be more links among nodes in the same community than among nodes in different communities. In other words, the link density within (across) communities should be higher (lower) than the overall network density [Eq. (1.3)]. This criterion can be mathematically formalized in many different ways. This is why we find many clustering techniques in the scientific literature.

This chapter offers a brief introduction to the problem and its most popular solutions. We start by introducing some basic elements: the main variables, classic definitions of communities, and high-level properties of partitions. Then we discuss two related problems,

---

[1] We encountered the term *clustering* when we introduced the clustering coefficient in Section 2.8; in this chapter we only use this term to refer to community structure.

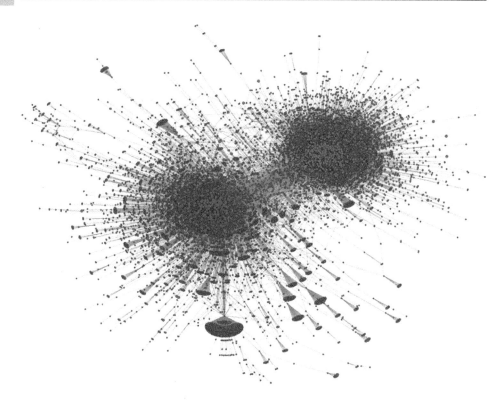

Retweet network of political hashtags on Twitter prior to the 2010 US election. Colored nodes are a sample of users classified as liberal (blue) or conservative (red). A link between two nodes indicates a retweet of one of the two corresponding users by the other.

network partitioning and data clustering, which have contributed many tools and techniques to the topic. Finally we present some of the widely adopted algorithms, along with standard procedures to test clustering techniques.

6.1 Basic Definitions

6.1.1 Community Variables

A community is typically a connected subnetwork. Consider the community illustrated by the green nodes in Figure 6.3. The magenta nodes are external but connected to the community, while some of the remaining nodes of the network are shown in black. The blue links connect the community to the rest of the network. The key community variables we will be using throughout this chapter are:

- The *internal* and *external degree* of a node in a community — the number of neighbors inside and outside the community, respectively. In Figure 6.3, the internal degree of a

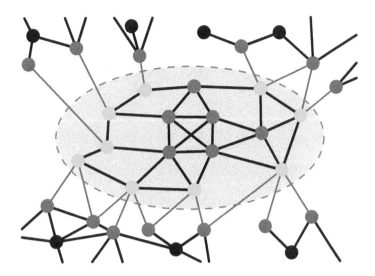

Fig. 6.3   Schematic picture of a community (inside the gray oval) and of its immediate neighbors. Reprinted from Fortunato and Hric (2016) with permission from Elsevier.

green node is the number of black links attached to it and the external degree is the number of blue links.

- The number of *internal links* to the community — the number of links connecting two nodes in the community (black links in the oval of Figure 6.3).
- The *community degree* — the sum of the degrees of the nodes in the community; the sum of the number of neighbors of each green node in Figure 6.3.
- The *internal link density* — the ratio between the number of internal links and the maximum number of links that could exist between any two community nodes. This is the same as the density defined in Chapter 1, but for the community subnetwork.

Let us formulate the definitions and notation for our community variables a bit more formally. Suppose you have a community C.

- The numbers of nodes and internal links in C are N_C and L_C, respectively.
- The *internal degree* k_i^{int} and the *external degree* k_i^{ext} of a node i with respect to community C are the numbers of links connecting i to nodes in C and to the rest of the network, respectively. Since every neighbor of i must be either inside or outside C, the degree of i is $k_i = k_i^{int} + k_i^{ext}$. If $k_i^{ext} = 0$ and $k_i^{int} > 0$, then i has neighbors only within C and is an *internal node* of C (dark green nodes in the figure). If $k_i^{ext} > 0$ and $k_i^{int} > 0$ for a node $i \in C$, then i has neighbors both inside and outside C and is a *boundary node* of C (bright green nodes in the figure). If $k_i^{int} = 0$, then the node is disjoint from C, as it has no neighbors inside the community (black nodes in the figure).
- The *internal link density* is given by

$$\delta_C^{int} = \frac{L_C}{\binom{N_C}{2}} = \frac{2L_C}{N_C(N_C - 1)}. \tag{6.1}$$

Note that this is equivalent to Eq. (1.3) for nodes and links internal to C, because we are assuming that the network is undirected; the maximum number of internal links that a community with N_C nodes may have is $\binom{N_C}{2}$.

- The *community degree*, or *volume*, is the sum of the degrees of the nodes in C:

$$k_C = \sum_{i \in C} k_i. \tag{6.2}$$

All definitions hold for undirected and unweighted networks. The extensions to weighted networks are straightforward, we just need to replace degrees with strengths. For instance, the internal degree of a node becomes the *internal strength*, which is the sum of the weights of the links connecting it to nodes in the community. For directed networks we need to distinguish between incoming and outgoing links. Extensions of the measures are fairly simple to implement, but their usefulness is unclear.

6.1.2 Community Definitions

Figure 6.1 is the traditional picture of network community structure. It emphasizes two things: (1) communities have high *cohesion* (i.e. they have many internal links, so the nodes stick together) and (2) communities have high *separation* (i.e. they are connected to each other by few links). Classic definitions of community-like structures are based on cohesion and on the interplay of cohesion and separation.

Definitions based on cohesion alone treat the community as a system of its own, disregarding the rest of the network. The most popular notion of community of this type is that of a *clique*, defined in Chapter 1 as a complete subnetwork, whose nodes are all connected to each other (Figure 6.4). However, in general, communities are not as dense as cliques. Moreover, all nodes have identical internal degree within a clique, whereas in real network communities some nodes are more important than others, as reflected in their heterogeneous linking patterns.

For a more useful definition of community, we should take into account both the internal cohesion of a candidate subnetwork and its separation from the rest of the network. A

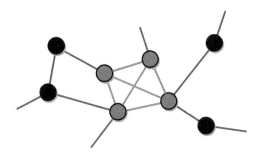

Fig. 6.4   Part of a network including a 4-node clique, identified by the blue nodes and the red links.

popular idea is that a community is a subnetwork such that *the number of internal links is larger than the number of external links*. This idea has inspired the following definitions:

- A *strong community* is a subnetwork such that each node has more neighbors in the subnetwork than in the rest of the network. In other words, the internal degree of each node in a strong community exceeds its external degree.
- A *weak community* is a subnetwork such that the sum of the internal degrees of all nodes exceeds the sum of their external degrees.

A strong community is necessarily also a weak community: if the inequality between internal and external degree holds for each node, then it must hold for the sum over all nodes. The converse is not generally true: if the inequality between internal and external degree holds for the sum, it may be violated for one or more nodes.

A drawback of these definitions is that they separate the community under consideration from the rest of the network, which is taken as a single object. But the rest of the network can in turn be partitioned into communities. If a subnetwork C is a proper community, one would expect each of its nodes to be more strongly attached to the other nodes in C than to the nodes in any other subnetwork of the partition. Such a concept has inspired less stringent definitions of strong and weak community:

- A *strong community* is a community such that each node has more neighbors inside it than in any other community.
- A *weak community* is a community such that the sum of the internal degrees of the nodes inside it exceeds the sum of their external degrees in each of the other communities. A node's external degree in a community other than its own is the number of neighbors in that community.

A strong (weak) community according to the earlier definition is necessarily also a strong (weak) community according to the less stringent definition. The converse is not true in general, as illustrated by the example in Figure 6.5. In particular, a subnetwork can be a strong community in the less stringent sense even though all of its nodes have internal degree smaller than their respective external degree.

As we have seen, traditional definitions of communities rely on counting links (internal and external) in various ways. But the number of links usually increases with the community size. Therefore, the comparisons between internal and external degrees of different communities are biased by their sizes. Ideally, we would like to compare *probabilities:* if nodes within a subnetwork are more likely to be connected than nodes in different subnetworks, we would call the subnetwork a community. Probabilities eliminate troublesome dependencies on community size. But how to determine link probabilities? For that, we need a model stating how links are formed in networks with community structure. Sections 6.3.4 and 6.4.1 present probabilistic models to define and detect communities.

Is a definition of community really necessary? Actually, most network clustering methods do not require a precise definition of community, as we will see in Section 6.3. However, defining criteria for communities beforehand can be useful when checking the reliability of the final results.

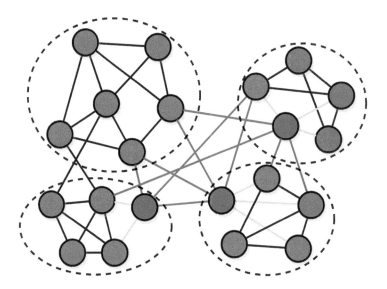

Fig. 6.5 Strong and weak communities. The four subnetworks enclosed in the dashed contours are weak communities according to both definitions we have given. They are also strong communities according to the less stringent definition, as the internal degree of each node exceeds the number of links joining the node with those of every other community. However, three of the subnetworks are not strong communities in the more stringent sense, because some nodes (in blue) have external degree larger than their internal degree (the internal and external links of these nodes are colored in yellow and magenta, respectively). Adapted from Fortunato and Hric (2016).

6.1.3 Partitions

A *partition* is a division, or grouping of a network into communities, such that each node belongs to only one community. The number of all possible partitions is called the *Bell number* and increases faster than exponentially with the number of nodes of the network. For instance, a network with 15 nodes has 1,382,958,545 possible partitions! Therefore, for networks with more than a handful of nodes, it is hopeless to choose the best partition of a network by going through all of them. Indeed, clustering algorithms typically explore only a tiny portion of the space of all partitions, where interesting solutions are most likely to be found.

The communities in many real networks *overlap* (i.e. they share some of their nodes). For instance, in social networks individuals can belong to different circles at the same time, like family, friends, and work colleagues. Figure 6.6 shows an example of a network with overlapping communities. A division of a network into overlapping communities is called a *cover*. The number of possible covers of a network is far higher than the already huge number of partitions, due to the many ways clusters can overlap.

Partitions can be *hierarchical* when the network has multiple levels of organization, at different scales. In this case, clusters display in turn community structure, with smaller communities inside, which may again contain smaller communities, and so on (Figure 6.7). For instance, in a collaboration network of employees of a multinational corporation, we

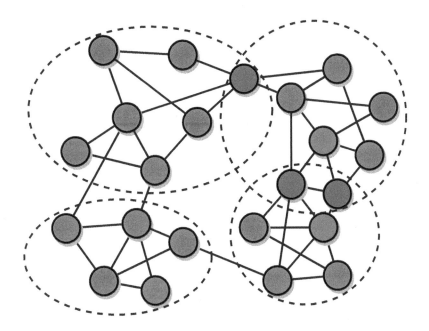

Fig. 6.6   Overlapping communities. We show a division of a network into four communities, enclosed by the dashed contours. Three of them share nodes, indicated in blue. Adapted from Fortunato and Hric (2016).

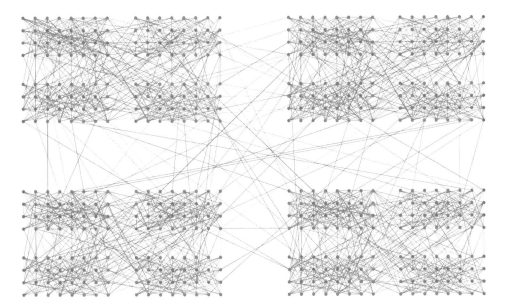

Fig. 6.7   A network with hierarchical communities. We can observe two hierarchical community structures: four large clusters with 128 nodes each and 16 small clusters with 32 nodes each. The smaller clusters are fully included within the larger ones.

would expect to distinguish clusters of employees working in the same branch, but within each branch we might see a further subdivision into departments. In such situations every level of the hierarchy has a meaning, and a good clustering method should be able to uncover all of them.

Partitions of real networks are often *heterogeneous*, in that some of the community properties may vary widely from one cluster to another. For instance, there is often a big difference in community size. In the Web, communities correspond roughly to pages or websites dealing with the same or similar topics. Since some topics are more general or popular than others, there are clusters with millions of Web pages along with clusters with just a few hundred or thousand pages. The cohesiveness is also very variable. If we measure it by the internal community link density, introduced in Section 6.1.1, we find that in several real networks this quantity spans orders of magnitude, implying that some clusters are much more cohesive than others. This may reflect a variable capacity of groups of nodes to "attract" and link to each other. But it might also be due to the dynamic character of the community formation process: some communities are fully developed because their nodes have been around for a sufficiently long time, while others may still be developing if many of their members have been introduced more recently.

6.2 Related Problems

6.2.1 Network Partitioning

We have seen that communities are usually well separated from each other. Identifying well-separated subnetworks is the goal of *network partitioning*. Here the focus is on the separation, regardless of how many links lie inside the subnetworks. Therefore network partitioning algorithms are not suitable to detect communities, in general. Nevertheless, some partitioning techniques are used to detect communities as well, typically in combination with other procedures. So it is useful to get acquainted with this problem.

Network partitioning is motivated by important practical problems. A classic example is parallel computing, where one aims at distributing tasks to processors, such that the number of communication links between groups of processors handling different tasks is low, to speed up the calculation. Network partitioning has been applied to problems in a large variety of domains: solution of partial differential equations and of sparse linear systems of equations, image processing, fluid dynamics, road networks, mobile communication networks, air traffic control, and more.

The partitioning problem consists of finding a division of a network into a given number of subnetworks, or clusters, of given sizes, such that the total number of links connecting nodes in different subnetworks is minimized. Figure 6.8 shows an example in which two clusters of equal size are desired; in this case the problem is also called *graph bisection*. The set of links joining the subnetworks to each other is called a *cut*, because their removal separates the clusters from each other, and their number is called

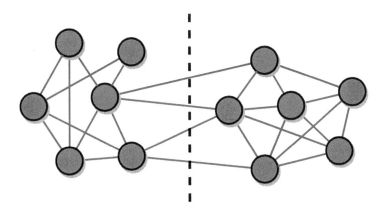

Fig. 6.8 Graph bisection. The network in the figure has 12 nodes. The goal is to divide it into two parts, having the same number of nodes, such that the number of links joining the parts is minimal. The solution is indicated by the vertical dashed line, which separates the two parts with a minimum cut size of four links.

the *cut size*. This is why the task is also known in the literature as the *minimum cut problem*.

   Why is it necessary to specify the number of clusters of the partition beforehand? After all, we could let the partitioning procedure find the optimal number. This is not an option though, because it gives a trivial solution: since we seek to minimize the cut size, the best possible partition has a single cluster, including the entire network, which yields a cut size of zero. The next question is why we must specify the sizes of the clusters. The reason again is to avoid trivial, non-informative solutions. For instance, if the network has one leaf (a node with degree one), the bipartition consisting of the leaf on one side and the rest of the network on the other has a cut size of one, as there is a single link separating the clusters. Such a solution cannot be beaten, but is not helpful. The main focus in network partitioning is the search for balanced solutions (i.e. partitions whose clusters have approximately the same size — as in Figure 6.8). In the case of bisection, if the network has an odd number of nodes, one cluster will have one node more than the other.

   One of the first and most popular methods to solve the graph bisection problem was the *Kernighan–Lin algorithm*. It is based on a very simple idea: given an initial bisection of the network, we swap pairs of nodes between the clusters, such as to obtain the greatest decrease of the cut size, while the size of the clusters does not change.

We start from an arbitrary partition P of the network into two clusters A and B. For instance, we can select half of the nodes at random and put them in one cluster, and the rest in the other cluster. Each iteration of the algorithm consists of the following steps:

1. For each pair of nodes i, j, with $i \in A$ and $j \in B$, compute the variation in cut size between the current partition and the one obtained by swapping i and j.
2. The pair of nodes i^* and j^* yielding the largest decrease in the cut size is selected and swapped. This pair of nodes is locked; they will not be touched again during this iteration.

3. Repeat steps 1 and 2 until no more swaps of unlocked nodes yield a decrease in the cut size. This yields a new partition P', that is used as a starting configuration for the next iteration.

The procedure ends when the cut size of partitions obtained after consecutive iterations is the same, meaning that the algorithm is unable to improve the result. The Kernighan–Lin algorithm can easily be extended to partitions with more than two clusters, by swapping nodes between pairs of clusters.

The solutions delivered by the Kernighan–Lin algorithm depend on the choice of the initial partitions. The poorer the quality of the initial partition — the higher its cut size — the worse the final solution and the longer the time to reach convergence. To obtain better outcomes, we can consider multiple random partitions and choose the one with the lowest cut size as the initial partition. Another limit is that Kernighan–Lin is a *greedy algorithm*, in that it tries to minimize the cut size at each step. A drawback of greedy strategies is that they are likely to get stuck in *local optima*, solutions with sub-optimal cut size such that any local swap leads to worse solutions. A more advanced version of the algorithm mitigates this limitation by occasionally swapping a pair of nodes that yields an increase in cut size. Accepting such moves may help escape sub-optimal solutions and more closely approach the absolute minimum of the cut size.

The Kernighan–Lin algorithm is widely applied as a post-processing technique, to improve partitions delivered by other methods. Such partitions can be used as starting points for the algorithm, which might return solutions with lower cut size.

Clusters identified via network partitioning are well separated but not necessarily cohesive, so they may not be good communities according to the widely accepted high-level definition we gave earlier. In addition, network partitioning requires the specification of the number of clusters to be found. While this is also a feature of a number of community detection methods, it would be preferable to be able to deduce this number directly from the data.

NetworkX has a function for bisectioning a network with the Kernighan–Lin algorithm:

```
# minimum cut bisection: returns a pair of sets of nodes
partition = nx.community.kernighan_lin_bisection(G)
```

6.2.2 Data Clustering

As we have seen, communities in networks tend to group nodes that are similar to each other somehow: papers or websites dealing with the same or related topics, people working in the same area or department, proteins having the same or similar cellular functions, and so on. Therefore, community detection is a special version of the much more general problem of *data clustering* (i.e. grouping data elements into clusters based on some notion of similarity, such that elements in the same cluster are more similar to each other than they are to elements in different clusters). Data clustering offers a valuable set of concepts and tools that are regularly used in network clustering as well.

There are two main classes of algorithms for data clustering: *hierarchical clustering*, which delivers a nested series of partitions, and *partitional clustering*, which yields only one partition. Hierarchical clustering is much more frequently adopted in network community detection than partitional clustering, so let us briefly discuss it here.

The main ingredient is a *similarity measure* between nodes. Such a measure may be derived from specific properties of the nodes. For instance, in a social network, it could indicate how close the profiles of two individuals are according to their interests. If the nodes can be embedded in a geometric space, which is often possible to do via suitable transformations, the distance between the points corresponding to a pair of nodes can be used as a dissimilarity measure for the nodes, so that points that are nearer to each other indicate more similar nodes. Alternatively, similarity measures can be derived from the structure of the network alone. A classic example is *structural equivalence*, which expresses the similarity between the neighborhoods of a pair of nodes.

The similarity S_{ij}^{SE} of a pair of nodes i and j via structural equivalence can be defined as

$$S_{ij}^{SE} = \frac{\text{number of neighbors shared by } i \text{ and } j}{\text{total number of nodes neighboring only } i, \text{ only } j, \text{ or both}} . \qquad (6.3)$$

For example, if the neighbors of i and j are (v_1, v_2, v_3) and (v_1, v_2, v_4, v_5), respectively, then $S_{ij}^{SE} = 2/5 = 0.4$, because there are two common neighbors (v_1 and v_2) out of five distinct neighbors in total (v_1, v_2, v_3, v_4, v_5). If a pair of nodes i,j have no common neighbors, then $S_{ij}^{SE} = 0$; if they have the same set of neighbors, then $S_{ij}^{SE} = 1$. We stress that i and j do not need to be neighbors: S_{ij}^{SE} can be computed for any pair of nodes.

The next step is to define the similarity between two *groups* of nodes. This can be done in several ways. The most popular approaches are *single linkage*, *complete linkage*, and *average linkage*. In these procedures, the similarity between two groups is determined via the similarity scores of pairs of nodes, where each pair consists of one node in each group.

Given a node similarity measure S and two groups of nodes G_1 and G_2 in a network, the similarity between G_1 and G_2 can be computed as follows. First, measure the similarity S_{ij} of all pairs of nodes (i,j), with $i \in G_1$ and $j \in G_2$. The similarity $S_{G_1 G_2}$ can be defined from these sets of pairwise similarities according to one of the following simple recipes.

- Single linkage uses the maximum pairwise similarity: $S_{G_1 G_2} = \max_{i,j} S_{ij}$.
- Complete linkage uses the minimum pairwise similarity: $S_{G_1 G_2} = \min_{i,j} S_{ij}$.
- Average linkage uses the average pairwise similarity: $S_{G_1 G_2} = \langle S_{ij} \rangle_{i,j}$.

Hierarchical clustering techniques are *agglomerative* if partitions are generated by iteratively merging groups of nodes, or *divisive* if partitions are generated by iteratively

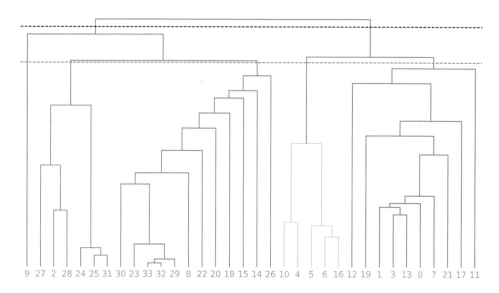

Fig. 6.9 Hierarchical clustering illustrated by a dendrogram of hierarchical partitions of Zachary's karate club network (see Section 6.4). Horizontal cuts single out partitions of the network. For instance, the black and red dashed lines correspond to partitions into two and five clusters, respectively. Each cluster includes the nodes "hanging" from one of the severed branches. Colors represent real and inferred clusters, as explained in the text.

splitting groups of nodes. Here we focus on agglomerative procedures, which are popular in the literature. Notable examples of divisive hierarchical clustering will be presented in Section 6.3.1.

Agglomerative hierarchical clustering starts from the trivial partition into N groups, where each group consists of a single node. At each step, the pair of groups with the largest similarity are merged. This is repeated until all nodes are in the same group. Since at each step the number of groups decreases by one, the procedure yields a series of N partitions, which can be represented via a *dendrogram*, or *hierarchical tree*. Figure 6.9 shows a dendrogram for partitions of a small network. At the bottom we have the leaves of the tree, which are the individual nodes, indicated by their labels. Going upwards, pairs of clusters are merged: each merge is illustrated by a horizontal line joining two vertical lines, each representing a cluster, whose nodes can be identified by following the vertical line all the way down. To single out one of the partitions we cut the dendrogram with a horizontal line, as shown in the figure. The vertical lines severed by the cut indicate the clusters of the partition. High cuts yield partitions into a small number of larger groups, whereas low cuts yield partitions into a large number of smaller groups. The partitions are hierarchical by construction: if we take any two partitions, every cluster of the one lying higher in the dendrogram is a merger of clusters of the lower one.

Hierarchical clustering has a number of important limitations. First, it delivers as many partitions as there are nodes, without providing a criterion that helps to choose which ones are meaningful for the network under study. Second, the results usually depend on the similarity measure and on the criterion adopted to compute the similarity of the groups. Finally, the algorithms are rather slow, and networks with millions or more nodes are out of reach.

6.3 Community Detection

There are many community detection methods. They are often classified into categories based on the strategy used to identify the clusters. Here, let us introduce four popular approaches: bridge removal, modularity optimization, label propagation, and stochastic block modeling. In the next section, we will show how to test the performance of a community detection algorithm.

6.3.1 Bridge Removal

Strictly speaking, a *bridge* is a link whose removal breaks a connected network into two parts. Here let us use this term more loosely and call any link joining two communities a bridge. If we were able to locate all bridges, we would have a natural way to detect the clusters: just remove the bridges, and the communities will be disconnected from each other! The problem would then be solved by finding the connected components of the resulting disconnected graph, which is trivial. This idea is behind the popular *Girvan–Newman algorithm* and several other community detection methods.

The key element of any algorithm based on bridge removal is a measure that allows us to identify the bridges. For the Girvan–Newman algorithm this measure is the link betweenness (Section 3.1.3). Recall that the link betweenness basically tells us how many shortest paths between pairs of nodes run through a link. We expect the link betweenness to be higher for bridges than for links inside the clusters: many shortest paths joining pairs of nodes in different communities necessarily run through bridges (Figure 6.10). In comparison, the betweenness of internal links is much lower, on average: due to the high density

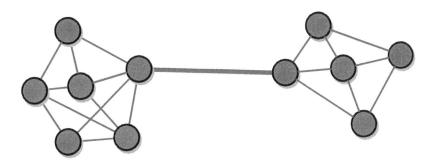

Fig. 6.10   Link betweenness and bridges. Link betweenness is usually high if the link is a bridge. In the figure, the link in the middle is a bridge; it has a much higher betweenness than all others, because every shortest path connecting nodes in the two communities must go through it.

of internal links, there are many alternative shortest paths crossing a community, so it is unlikely that any one of them is a preferential route.

The Girvan–Newman algorithm iteratively identifies and removes the link with largest betweenness, which leads to a progressive breakup of the network into disconnected pieces.

> We start by calculating the betweenness for all links. Then, each iteration of the algorithm consists of two steps:
>
> 1. Remove the link with largest betweenness; in case of ties, one of them is picked at random.
> 2. Recalculate the betweenness of the remaining links.
>
> The procedure ends when all links are removed and the nodes are isolated.

In each iteration, the betweenness of all links must be recalculated. This is critical to obtain meaningful results, but makes the method very slow. On networks with strong community structure, which quickly break into disconnected communities, the recalculation step needs to be performed only within the connected component including the last removed link, as the betweenness of all other links remains the same. This can significantly reduce the computational burden of betweenness recalculation.

The Girvan–Newman algorithm delivers a set of N partitions, from the one consisting of a single cluster including the entire network to the one where each node is a singleton and forms its own community. In each partition, the clusters correspond to the connected components of the network. Whenever the removal of a link splits one network component, the number of clusters increases by one. All partitions are hierarchical because the removal of links breaks clusters into smaller pieces, which are in turn split into smaller pieces, and so on. This is an example of divisive hierarchical clustering, the opposite of agglomerative hierarchical clustering (Section 6.2.2). As for agglomerative hierarchical clustering, the full set of partitions delivered by the method can be represented by a dendrogram. In Figure 6.9 we show the partitions detected by the Girvan–Newman method on Zachary's karate club network, which represents social interactions between the members of a karate club. This is a classic benchmark in community detection, that we describe in Section 6.4. The network has a natural partition into two communities, indicated by the different colors of the node labels at the bottom of the dendrogram. The partition into two clusters found by the algorithm (corresponding to the black horizontal cut) coincides with the natural division of the network, with the exceptions of nodes **2** and **8**, which are misclassified.

NetworkX has a function for the Girvan–Newman algorithm:

```
# returns a list of hierarchical partitions
partitions = nx.community.girvan_newman(G)
```

Because the method is quite slow, it is not practical for large networks with, say, more than 10,000 nodes. The bottleneck is the recalculation of the link betweenness. Faster variants have been proposed to get around this problem. For instance, instead of computing the betweenness exactly by using all possible pairs of nodes to determine the number of shortest paths crossing a link, one could obtain approximate values of the betweenness scores by using only a sample of the node pairs. What matters is the ranking of the links and not their exact betweenness values. Scholars have also proposed alternative measures to identify bridges, which are not as expensive to compute as betweenness.

The other major drawback of the algorithm is common to every hierarchical clustering technique (Section 6.2.2): since there are as many partitions as there are nodes, which ones are meaningful, if any? To answer this question, in the next section we introduce a measure expressing the quality of a partition. One can use such a measure to select the best partition from a dendrogram.

6.3.2 Modularity Optimization

How can we tell how "good" a partition is? A natural approach is to measure how community-like the subnetworks of the partition are. For instance, we could compute the internal link density of each subnetwork and see if it is sufficiently high. But such a strategy might lead to wrong results. Consider for example a random network (Section 5.1): we do not expect to find communities in it, because nodes are connected to each other randomly, so there is no group of nodes that prefer to link to each other over linking to nodes outside the group. This is a relatively well-established principle of community detection: *random networks do not have communities!* Such a statement is independent of link density. Consequently, a random subnetwork does not make a good community, irrespective of its internal link density. We need a better way to measure the quality of a community partition.

The *modularity* of a partition evaluates the communities not in absolute terms, but with respect to a *random baseline*. It does so by discounting the internal links that could be attributed to a randomized version of the original network. In simple terms, the modularity is the difference between the number of links internal to all the clusters and the value of this number expected in a randomized network. The baseline adopted in this definition is the degree-preserving randomization discussed in Section 5.3 — networks with the same number of nodes and where each node maintains the degree it has in the original network. If the original network is a random graph, it will have similar features as its randomizations, and its modularity will be low. Specifically, if the number of internal links of each cluster is close to its expected value in random versions of the network, the community structure of the network is compatible with that of a random network with the same degree sequence, and the modularity is low. In contrast, if the number of links within the clusters is much larger than its expected random value, it is unlikely for such a concentration of internal links to be the result of a random process, and the modularity can reach high values (Figure 6.11). Box 6.1 presents a formal definition of modularity in undirected, unweighted networks. Given a partition of such a network into a set of communities, the modularity value [Eq. (6.4)] can be calculated with the following NetworkX function:

```
# returns the modularity of the input partition
modularity = nx.community.quality.modularity(G, partition)
```

Box 6.1   Modularity

The modularity of a partition in an undirected, unweighted network is defined as

$$Q = \frac{1}{L} \sum_C \left(L_C - \frac{k_C^2}{4L} \right),$$
(6.4)

where the sum runs over all clusters of the partition, L_C is the number of internal links in cluster C, k_C is the total degree of the nodes in C [Eq. (6.2)], and L is the number of links in the network.

Each element of the sum is the difference between the number of internal links in cluster C and its expected number in degree-preserving randomizations of the network. To calculate the expected number, consider that random links are formed by matching pairs of stubs chosen at random (Section 5.3). The total number of stubs attached to nodes in C is the cluster's total degree, which is k_C in every randomization, because each node preserves its degree. The probability of selecting one of those stubs at random is $k_C/2L$, because $2L$ is the total number of stubs of the network (each link yields two stubs). For a random link to connect two nodes in the same cluster C, two stubs must be selected from C. To a close approximation, the probability of picking a pair of stubs from C at random is the product of the probabilities of selecting each one: $\frac{k_C}{2L} \cdot \frac{k_C}{2L} = \frac{k_C^2}{4L^2}$. (This scenario is similar to calculating the probability of drawing heads twice in two independent tosses of a fair coin as $\frac{1}{2} \cdot \frac{1}{2} = \frac{1}{4}$.) Finally, since there are L links in total and each joins two nodes in C with probability $k_C^2/4L^2$, the expected number of internal links is $L \cdot k_C^2/4L^2 = k_C^2/4L$.

The partition of any network in a single cluster yields $Q = 0$. This is because the sum in Eq. (6.4) reduces to a single term, $L_C = L$, and $k_C = 2L$ is the total degree of the network. Further, $Q < 1$ for any partition of any network because it cannot be larger than $(\sum_C L_C)/L$, which is at most one when all links are internal. Modularity can take negative values. Consider the partition into N singletons: the first term in each summand is zero, because there are no links joining a node to itself, thus Q is a sum of negative numbers. For most networks, the modularity has a non-trivial maximum: $0 < Q_{max} < 1$.

The definition can be extended directly to partitions in directed and weighted networks.

The expression of modularity for partitions of directed networks is

$$Q_d = \frac{1}{L} \sum_C \left(L_C - \frac{k_C^{in} k_C^{out}}{L} \right),$$
(6.5)

where L_C is the total number of directed links inside cluster C and k_C^{in}, k_C^{out} are the total in-degree and out-degree of the nodes in C, respectively.

For weighted networks we have

$$Q_w = \frac{1}{W} \sum_C \left(W_C - \frac{s_C^2}{4W} \right),$$
(6.6)

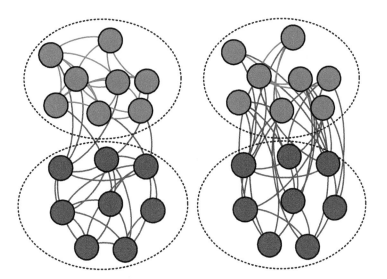

Network modularity. The network on the left has a visible community structure, with two clusters whose nodes are highlighted in blue and red, and consequently high modularity. The picture on the right shows a degree-preserving randomization of the network. In the randomized network there are fewer internal (blue and red) links compared to the original network, and more (gray) links between the subnetworks: the randomization process has destroyed the community structure of the network. The modularity of the same partition is therefore smaller. Reproduced from Fortunato and Hric (2016) with permission from Elsevier.

where W is the total weight of the links of the network, W_C the total weight of the internal links in C, and s_C the total strength of the nodes in C [i.e. the sum of the strengths of the nodes in C; Eq. (1.8)].

For networks whose links are both directed and weighted, we have

$$Q_{dw} = \frac{1}{W} \sum_{C} \left(W_C - \frac{s_C^{in} s_C^{out}}{W} \right),$$ (6.7)

where s_C^{in}, s_C^{out} are the total in-strength and out-strength of the nodes in C, respectively.

Modularity was originally introduced to rank the partitions delivered by the Girvan–Newman algorithm, so one can choose the best one. For instance, the largest modularity value in the dendrogram of Figure 6.9 is obtained by partitioning Zachary's karate club network into the five clusters corresponding to the dashed red cut and indicated by the colors of the vertical lines. But since modularity measures the quality of any partition, we are not limited to using it in conjunction with other techniques. All we have to do is search for the partition with the largest possible modularity. This is the idea behind *modularity optimization*, a large class of network clustering algorithms. Recall from Section 6.1.3 that the number of possible partitions of a network is huge, so it is impossible to carry out a complete exploration of the space of partitions, even for small networks. Good algorithms usually constrain their search to a tiny subset of partitions.

A simple modularity maximization technique is an agglomerative algorithm that starts from the partition in which each node is its own cluster, then iteratively merges the pair of clusters that yields the largest increase in modularity. The method explores a hierarchy of partitions, represented by a dendrogram. The initial partition into singletons has negative modularity, then the modularity increases steadily until a positive peak is reached, and finally it goes down until it hits zero when all nodes are in the same community. The partition corresponding to the peak value is the best solution found by the algorithm. This method is greedy, in that it tries to maximize the modularity at each step. As such, it is likely to get stuck on solutions with sub-optimal modularity. The algorithm also tends to generate unbalanced partitions, with some clusters much larger than others. Such unbalanced partitions are often unrealistic and slow down the method considerably. Simple modifications to the algorithm, such as merging groups of comparable size or more than two groups at a time, have proven effective at mitigating this problem.

NetworkX has a function for a fast version of greedy modularity optimization:

```
# returns the maximum modularity partition
partition = nx.community.greedy_modularity_communities(G)
```

The most popular modularity optimization method is the *Louvain algorithm*. It is another agglomerative procedure in which communities are iteratively turned into supernodes, as illustrated schematically in Figure 6.12.

The algorithm starts again from the partition into singletons. Each iteration consists of two steps.

1. Loop over the nodes: each node is put into the community of the neighbor that yields the largest modularity increase ΔQ with respect to the current partition. All nodes are revisited over and over until it is no longer possible to increase Q by moving a node to a different community.
2. The network is transformed into a weighted supernetwork, where each community in the partition from step 1 is replaced by a supernode, links between supernodes carry a weight corresponding to the number of links joining nodes in the corresponding groups, and links joining nodes in the same community are represented as a self-loop from the corresponding supernode to itself, with weight equal to the number of internal links.

Since we ultimately care about clustering the actual nodes and not the supernodes, modularity is always computed with respect to the original network. The procedure stops when no further grouping of the clusters in the current partition increases the modularity, and returns the partition with the largest modularity.

The Louvain algorithm is a greedy technique, so it usually fails to find a partition that closely approximates the optimal modularity. Moreover, the result depends on the order in which the nodes are visited. On the plus side, the algorithm is very fast because, after

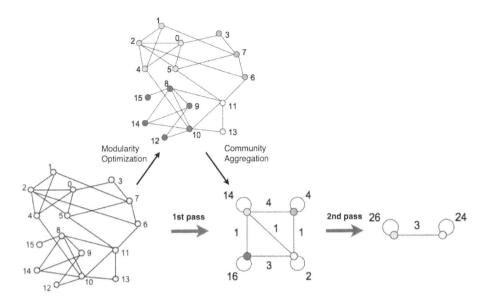

Fig. 6.12   Louvain algorithm. We show two iterations of the method, starting from the graph on the left. Each iteration consists of two steps: first, every node is assigned to the neighboring community yielding the largest (positive) modularity gain, until modularity cannot be increased further. Next, the network is transformed into a smaller weighted graph by turning clusters into supernodes, each set of links between two distinct communities into a single weighted link between the corresponding supernodes, and internal links in each community into a self-loop of the corresponding supernode. The weights on the superlinks allow us to more quickly compute the variation of modularity for the partitions of the original network corresponding to the merger of groups represented by the supernodes. Figure reprinted with permission from Blondel *et al.* (2008), copyright (2008) by the Institute of Physics Publishing.

the first iteration, successive transformations shrink the network very quickly and typically only a handful of partitions are generated. The smaller networks allow for swift calculations. Therefore the method is widely used in practical applications; for example, the clusters shown in Figure 0.2(b) were found in this way. The Louvain algorithm can be used to detect communities in large networks with many millions of nodes and links.

NetworkX does not yet have an implementation of the Louvain algorithm at the time of writing this book. One is available by importing the `community` module:[2]

```
# download community module at
# github.com/taynaud/python-louvain
import community
# returns the partition with largest modularity
partition_dict = community.best_partition(G)
```

[2] python-louvain.readthedocs.io

While modularity methods are popular, the measure has important limits that undermine its usefulness in practical applications. For instance, the maximum modularity tends to be larger on larger networks. Therefore, it cannot be used to compare the quality of partitions in different systems. Furthermore, the maximum modularity of partitions of random networks may attain fairly large values. This may sound surprising, as the measure is defined with respect to a random baseline, so if a network is itself random we would expect small deviations from the baseline. But the measure simply subtracts the expected number of internal links for each community from the actual number [Eq. (6.4)]; it does not take into account random fluctuations around the expected value, which may inflate the modularity. Finally, the maximum modularity does not necessarily correspond to the best partition. This is because the measure has an intrinsic *resolution limit* that prevents it from detecting small communities.

Specifically, the resolution limit depends on the total number of links in the network: communities whose degree is smaller than $\sqrt{2L}$ are virtually invisible for the method and may be merged with other clusters.

Figure 6.13 shows an extreme consequence of this problem. Since cliques are the most cohesive subnetworks we can get, as all internal links are present, the network in the figure has a natural partition into 16 communities, corresponding to the cliques. However, there are partitions with larger modularity, such as the eight clusters indicated by the dashed contours in the figure. A way to get around this problem is to tune the resolution of the method by introducing a parameter in the definition of modularity [Eq. (6.4)], whose value sets the

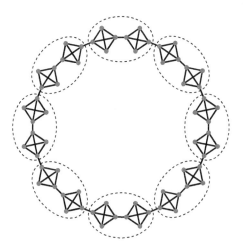

Fig. 6.13 Resolution limit of modularity optimization. A network is made of cliques of four nodes, forming a ring-like structure, with each clique joined to two others by single links. We would expect modularity to peak for the partition whose communities are the cliques. This partition has modularity $Q \approx 0.795$. However, it turns out that the partition combining pairs of cliques (indicated by the dashed contours) has higher modularity, $Q \approx 0.804$. Reprinted from Fortunato and Hric (2016) with permission from Elsevier.

scale of the detected communities, from very small to very large sizes. This strategy, called *multiresolution modularity optimization*, is computationally demanding because modularity has to be optimized for multiple choices of the parameter. Furthermore, a criterion is needed to decide which value of the resolution parameter is most suitable for a given network. Despite these shortcomings, multiresolution modularity optimization is widely used in applications.

6.3.3 Label Propagation

The *label propagation algorithm* is a simple and fast community detection method based on the idea that neighbors usually belong to the same community. This amounts to saying that most links are internal, yielding cohesive and separated communities as discussed in Section 6.1.2. At each step, the algorithm inspects each node and assigns it to the community of the majority of its neighbors. The procedure eventually converges to a stable partition, where every node has the same community membership as the majority of its neighbors.

> The label propagation method starts from the partition of singletons. Each node is given a different label. The procedure follows two iterative steps.
>
> 1. A sweep is performed over all nodes, in random order: each node takes the label shared by the majority of its neighbors. If there is no unique majority, one of the majority labels is picked at random.
> 2. If every node has the majority label of its neighbors (*stationary state*), stop. Else, repeat step 1.
>
> Communities are defined as groups of nodes having identical labels in the stationary state.

During this process, labels propagate across the network: most labels disappear, others dominate. Since the partition of the network changes during each sweep, multiple sweeps are needed to reach the stationary state. However, the algorithm typically converges after a small number of iterations, fairly independent of the network size.

In the final partition, each node has more neighbors in its own community than in any other. So each cluster is a strong community according to the less stringent definition given in Section 6.1.2. A problem is that the algorithm does not deliver a unique solution; the outcome depends on the order in which the nodes are visited in each sweep, which is set to be random to avoid bias but by the same token varies depending on the sequence of random numbers used in the calculation. Different partitions are also the result of the many ties encountered along the process, which can be broken in different ways depending again on the sequence of random numbers. Despite these instabilities, the partitions found by label propagation in real networks tend to be similar to each other. For more robust results, one can combine the solutions obtained from different runs of the procedure.

The strength of the method is the fact that it does not need any information on the number and size of the communities. It does not have any parameter, either. The technique is simple to implement and very fast: networks with millions of nodes and links can be partitioned this way. If community labels are known for some of the nodes, they can be used as seeds in the initial partition. As an example, this method was used to assign colors to the nodes in the Twitter diffusion network shown in Figure 0.3.

NetworkX has a function for the label propagation algorithm:

```
partition = nx.community.asyn_lpa_communities(G)
```

6.3.4 Stochastic Block Modeling

Suppose you have a network and know that it was generated by some model, defined by a set of parameters. Once the values of the parameters are fixed, the model generates a class of networks. If the parameters are unknown, we can ask for which values the model networks most closely resemble our graph. This is similar to when we fit, say, a straight line to a set of data points and infer the slope parameter. For instance, if we know that our network is a random graph, we might ask which value of the link probability produces graphs whose structure is most similar to that of our network.

If we are interested in uncovering the community structure of the network, we can think of models that generate networks with communities. This way, once the best-fitting model is found, the communities planted in the model are the best approximation of the clusters we wish to detect. The most widely adopted model of networks with communities is the *stochastic block model* (SBM). The basic idea is that nodes are divided into groups and the probability that two nodes are connected is determined by the groups to which they belong.

Formally, let the N nodes of a network be divided into q groups $1, \ldots, q$. Node i is in group g_i. The probability that nodes i and j are connected depends exclusively on their group memberships: $P(i \leftrightarrow j) = p_{g_i g_j}$. Therefore, for any pair of groups g_1 and g_2, the connection probability between any node in g_1 and any node in g_2 is identical. The probabilities form a $q \times q$ matrix, called the *stochastic block matrix* (Figure 6.14). For directed graphs, the matrix is in general asymmetric. Here we focus on undirected

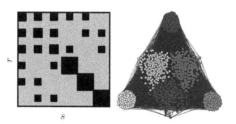

Fig. 6.14 Stochastic block models. (Left) Stochastic block matrix of an SBM with six blocks. The size of the squares is proportional to the values of the link probabilities between the corresponding groups. (Right) Realization of a network with 1000 nodes produced by using the link probabilities of the matrix. The groups are identified by the colors. Figure reprinted from Peixoto (2012) with permission from the American Physical Society.

networks, so that the matrix is symmetric. The diagonal elements p_{gg} $(g = 1, \ldots, q)$ of the stochastic block matrix are the probabilities that nodes in block g are neighbors, whereas the off-diagonal elements give the link probabilities among different blocks.

If the probability of connecting nodes within the same group is higher than that of links between nodes in different groups, then the model yields networks with cohesive and separated communities. But the model can also generate various other types of group structure.

If $\forall r, s = 1, \ldots, q$ with $r \neq s$ we have $p_{rr} > p_{rs}$, then we recover community structure. For $p_{rr} < p_{rs}$ we have *disassortative structure*, as links are more likely between blocks than inside them. In the special case $p_{rr} = 0$ $\forall r$, we get *multipartite networks*, as there are links only between the blocks. If $q = 2$ and $p_{11} \gg p_{12} \gg p_{22}$, we have *core–periphery structure*: the nodes in the first block (core) are relatively well connected amongst themselves as well as to a peripheral set of nodes that interact very little among themselves. Finally, if all probabilities are equal ($\forall r, s : p_{rs} = p$), we recover the classic random network: any two nodes have identical probability of being connected, hence there is no group structure.

With the model defined, we have to fit it to our network. The standard procedure is to maximize the likelihood that, for a given partition of the network, an SBM reproduces the placement of links between the nodes. This likelihood can be computed analytically; it tells us how well the SBM with that given partition mimics our network. The last step is to find the partition that yields the largest value of the likelihood. Box 6.2 expresses the likelihood and presents a greedy algorithm to find a partition that maximizes it. Like all greedy techniques, the method provides a sub-optimal solution. To improve the result, it helps to run the algorithm several times with different random initial conditions and select the partition with highest likelihood across all runs.

The most important limit of this approach is the need to specify the number of groups beforehand, which is usually unknown for real networks. This is because, as it turns out, a straight maximization of the likelihood over the whole set of possible partitions yields a trivial division into singletons. Fortunately there are ways to estimate the number of clusters, either beforehand or through more refined SBMs.

6.4 Method Evaluation

How do we know that a community detection method is "good"? How can we tell which of two methods is better than the other? Such questions are tricky because in general there is no ground truth about the right way to partition a network. A common way to evaluate an algorithm is to check if it is able to find communities in *benchmark graphs* (i.e.

Box 6.2   **Fitting a Stochastic Block Model to a Network**

The standard SBM model does a poor job of describing the group structure of most real networks because it ignores degree heterogeneity. Therefore we consider the *degree-corrected stochastic block model* (DCSBM), in which the degrees of the nodes match the actual degrees in the network. The probability that a network G is reproduced by the DCSBM based on a given partition g of G's nodes into q groups is expressed by the *log-likelihood*

$$\mathcal{L}(G|g) = \sum_{r,s=1}^{q} L_{rs} \log \left(\frac{L_{rs}}{k_r k_s} \right), \tag{6.8}$$

where L_{rs} is the number of links running from group r to group s, k_r (k_s) is the sum of the degrees of the nodes in r (s), and the sum runs over all pairs of groups in g (including when $r = s$).

The maximization of the likelihood in Eq. (6.8) for a partition into q groups can be achieved via a simple greedy technique. The starting point is a random partition into q clusters. Each iteration of the algorithm consists of two steps:

1. Repeatedly move a node from one group to another, selecting at each step the move that will most increase the likelihood (or least decrease it, if no increase is possible), under the constraint that each node may be moved only once.
2. When all nodes have been moved, inspect the partitions through which the system passed from start to end of the procedure in step 1 and select the one with the highest likelihood.

The algorithm stops when the likelihood for two consecutive iterations is the same (i.e. it cannot be increased any further).

networks known to have a "natural" community structure). There are two classes of benchmarks: (i) artificial networks generated via some model and (ii) real networks in which the communities are suggested by the history of the system or by attributes of the nodes.

6.4.1 Artificial Benchmarks

Stochastic block models (Section 6.3.4) are often used to generate artificial benchmark graphs. NetworkX has a function to do this:

```
# network with communities with sizes in the list S
# and stochastic block matrix P
G = nx.generators.stochastic_block_model(S, P)
```

A special version of SBM is the *planted partition model*. It is a simplified version of the original SBM, with only two link probabilities: the probability that two nodes in the same community are connected and the probability that two nodes in different communities are connected.

In Section 6.3.4 we have seen that the standard SBM with q groups is characterized by a $q \times q$ stochastic block matrix whose element p_{rs} expresses the probability of having a link between any node in group r and any node in group s. In the planted partition model, $p_{rs} = p_{int}$ for $r = s$ and $p_{rs} = p_{ext}$ for $r \neq s$. If $p_{int} > p_{ext}$, there is a higher chance of two nodes being connected if they are in the same group than if they are in different groups, implying that the groups are communities. The model further assumes that all communities have identical size N/q. Given the values of p_{int}, p_{ext}, and q, we can compute the expected internal and external degrees of a node: $\langle k^{int} \rangle = p_{int} \left(\frac{N}{q} - 1 \right)$ and $\langle k^{ext} \rangle = p_{ext} \frac{N}{q}(q-1)$, as each of the other $\frac{N}{q} - 1$ nodes in the group of any node i has equal probability p_{int} of becoming a neighbor of i, and each of the $\frac{N}{q}(q-1)$ nodes in the other groups has equal probability p_{ext} of becoming a neighbor of i. The expected (total) degree is $\langle k \rangle = \langle k^{int} \rangle + \langle k^{ext} \rangle = p_{int} \left(\frac{N}{q} - 1 \right) + p_{ext} \frac{N}{q}(q-1)$.

NetworkX has a function to generate networks according to the planted partition model:

```
# network with q communities of nc nodes each
# and link probabilities p_int and p_ext
G = nx.generators.planted_partition_graph(q, nc, p_int, p_ext)
```

A specific implementation of the planted partition model, in which the network size, degree of the nodes, and number and size of the communities are set to particular values, is known as the *GN benchmark* and has been used by the scientific community as a standard validation tool for a long time.

To derive the GN benchmark we set $N = 128$, $q = 4$, and $\langle k \rangle = 16$. This implies that $31 p_{int} + 96 p_{ext} = 16$, so that p_{int} and p_{ext} are not independent parameters. Once we fix p_{int}, the value of p_{ext} is determined by this relation. Knowing p_{int} and p_{ext}, the network is constructed with a procedure similar to the one adopted for Erdős–Rényi random graphs (Section 5.1): we go over all pairs of nodes and connect each with probability p_{int} or p_{ext}, depending on whether or not the nodes are in the same community.

The higher the external degree and the lower the internal degree, the more difficult it is to detect the communities. Figure 6.15 shows three GN benchmark networks of increasing difficulty.

For low values of $\langle k^{ext} \rangle$, the communities are well separated and most algorithms detect them without problem. Performance declines with increasing $\langle k^{ext} \rangle$: a growing number of nodes end up having comparable numbers of neighbors in different groups and can be misclassified. As long as $p_{int} > p_{ext}$, a good algorithm should in theory be able to recognize the communities. When $p_{int} = p_{ext} = 16/127$, $\langle k^{ext} \rangle \approx 12$; communities should then be detectable for $\langle k^{ext} \rangle$ below this value. Instead, it turns out that the detectability threshold is quite a bit lower, around nine, due to random fluctuations in the placement of the links. For $9 \leq \langle k^{ext} \rangle \leq 12$, these fluctuations make the networks basically indistinguishable from random graphs.

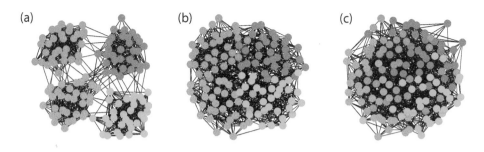

Fig. 6.15 Three GN benchmark networks with expected degrees (a) $\langle k^{ext} \rangle = 1$, $\langle k^{int} \rangle = 15$, (b) $\langle k^{ext} \rangle = 5$, $\langle k^{int} \rangle = 11$, and (c) $\langle k^{ext} \rangle = \langle k^{int} \rangle = 8$. The four groups in network (c) are hardly distinguishable; in this case, community detection methods fail to assign many nodes to the correct groups.

Despite its usefulness, the GN benchmark is not a good proxy for real networks with community structure. One problem is that all nodes have approximately equal degree, whereas the degree distribution of real networks is usually highly heterogeneous (Section 3.2). Another limit is that communities in real networks usually have quite different sizes, whereas benchmarks based on the planted partition model yield groups of identical size. The more advanced *LFR benchmark* produces networks having heavy-tailed distributions of degree and community size, as illustrated in Figure 6.16. The LFR benchmark is now used regularly to evaluate community detection algorithms.

When evaluating community detection methods, *negative tests* are important as well. To this end one can use networks without community structure. Random networks are good examples; an algorithm that detects communities in them is unlikely to be reliable in applications. The meaningful partitions that we expect to get in those cases are the one into singletons and the one where the entire network is a single cluster. Any other division would signal the method's inability to distinguish actual communities from subnetworks with high concentrations of links generated by random fluctuations.

6.4.2 Real Benchmarks

The most famous example of a real network with communities is *Zachary's karate club network*, illustrated in Figure 6.17. It has 34 nodes, the members of a karate club in the United States, who were monitored over a period of 3 years. Links connect individuals interacting outside the activities of the club. At one point in time, a conflict between the instructor and the president of the karate club led to a split of the club into two separate groups, whose members supported the instructor and the president, respectively. The club groups make sense based on the network structure: most members are connected to one of the two hubs, revealing their close association with the president or the instructor. A reliable clustering technique should be able to recognize the bipartition. In fact, Zachary's karate club network does not pose a tough challenge for community detection algorithms, many of which are capable of correctly classifying the nodes.

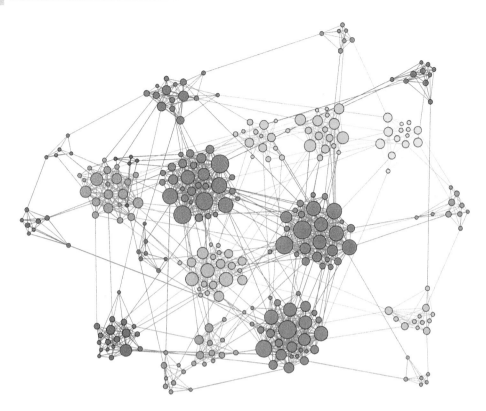

Fig. 6.16   LFR benchmark. Node size is proportional to degree. Node degree and community size are broadly distributed to account for the heterogeneity observed in most real networks with community structure.

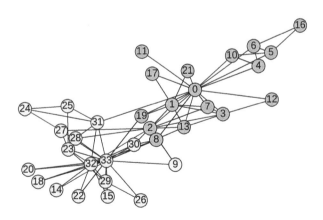

Fig. 6.17   Zachary's karate club network. The colors indicate the followers of the club's instructor (node 0) and its president (33), which ended up forming two separate groups.

NetworkX has a function that returns Zachary's karate club network:

```
G = nx.karate_club_graph()
```

Many other networks, whose nodes can be classified based on their attributes, are available for testing community detection algorithms. For instance, in many social networks there are groups that users can decide to join; in citation networks papers can be grouped according to their publication venues; Internet routers can be categorized by country; and so on. Such groups do not always match the communities found by clustering methods, raising the question of whether structural clusters have to correspond to attribute-based groups. The answer depends on the network and the attributes. Detecting communities in networks is a way to uncover hidden attributes that determine what the graph looks like. So, if nodes with equal or similar attributes are strongly linked to each other, the attributes can be revealed through community detection. If instead the attributes do not play a role in the build-up of the network, they remain invisible to clustering methods.

6.4.3 Partition Similarity

The last ingredient required by a validation procedure is a *partition similarity* measure, telling how similar the outcome of the algorithm is to the natural partition of the network adopted as a benchmark. There are many partition similarity measures. They can be classified according to the criteria used to estimate the similarity.

A popular measure is the *fraction of correctly detected nodes*. A node is correctly classified if it, and at least half of the other nodes in the same community in the detected partition, are in the same community in the benchmark partition. Note that if the detected partition has communities obtained by merging two or more groups of the benchmark partition, all the nodes of those clusters are considered incorrectly classified by this measure. The number of correctly classified nodes is then divided by the number of nodes in the network, yielding a fraction between zero and one. A problem of this measure is that the recipe to label nodes as correctly or incorrectly classified is somewhat arbitrary.

A better approach is to estimate the similarity between two partitions by computing, given one of the two, the additional amount of information needed to infer the other. This amounts to the information about the nodes that must be moved between clusters to transition from one partition to the other. If the partitions are similar, little information is needed to go from one to the other. The more extra information is needed, the less similar the partitions. Box 6.3 presents the *normalized mutual information*, a similarity measure based on information. Though widely adopted, normalized mutual information has some limitations. Detected partitions with more clusters may yield larger values, even though they are not necessarily closer to the benchmark partition. This may give wrong perceptions about the relative performance of algorithms. Other partition similarity measures can be used, but any measure has advantages and disadvantages.

Box 6.3	Normalized Mutual Information

Some measures of similarity between partitions use concepts borrowed from information theory. The probability that a randomly chosen node belongs to cluster x of partition X is given by $P(x) = N_x/N$, where N_x is the size of x. The probability that a randomly chosen node belongs to both cluster x of partition X and cluster y of partition Y is $P(x, y) = N_{xy}/N$, where N_{xy} is the number of nodes that clusters x and y have in common. The *normalized mutual information* of partitions X and Y is defined as

$$\text{NMI}(X, Y) = \frac{2H(X) - 2H(X|Y)}{H(X) + H(Y)}, \tag{6.9}$$

where $H(X) = -\sum_x P(x) \log P(x)$ is the *Shannon entropy* of X and $H(X|Y) = \sum_{x,y} P(x, y) \log[P(y)/P(x, y)]$ is the *conditional entropy* of X given Y. The sums are over all clusters x of partition X and over all pairs of clusters x and y of X and Y. NMI $= 1$ if and only if the partitions are identical, whereas it has an expected value of zero if they are independent, as for example when two random partitions are compared.

6.5 Summary

Communities play a key role in the structure and function of networks. They reveal similarities between nodes, show how the network is organized, allow us to uncover the role of the nodes both in their community and in the network as a whole, and affect the dynamics of processes running on the network. This is why community detection is a central problem in network analysis. Here is what we have learned in this chapter.

1. Communities are not well-defined objects. At a high level, they are cohesive and well-separated subnetworks, in that there are many links inside them, and not so many between them. Many clustering algorithms do not require an exact definition of community.
2. The number of possible partitions of a network into communities is enormous, even for a small graph, so we cannot search them all.
3. Random networks do not have communities. They can be used to test whether community detection algorithms can distinguish signal from noise.
4. Network partitioning searches for well-separated subnetworks. These are not necessarily cohesive, so they may or may not correspond to communities. Despite this limitation and the fact that the number of clusters must be specified as input, network partitioning tools can be useful for community detection.
5. Hierarchical clustering groups nodes based on their similarity. It is frequently used but it has a number of drawbacks, most notably the lack of a criterion to select meaningful partitions out of the full hierarchy (dendrogram) delivered by the procedure.
6. Bridge removal consists of erasing links between communities, so that the latter can be disconnected and identified. Like other hierarchical clustering approaches, bridge

removal has the problem of not being able to rank the hierarchical partitions it finds, unless an additional criterion is provided.

7. Modularity optimization searches for the partition with the largest modularity score. Modularity measures the quality of a partition by comparing the number of internal links with those expected in randomized versions of the network. The larger the score, the less random and therefore the more significant the clusters. The Louvain method is a greedy modularity optimization technique that is used extensively in applications thanks to its speed. One limitation of modularity optimization is that networks without group structure may have partitions with fairly large modularity. Another is that small communities may not be found.

8. The label propagation method assigns nodes to communities such that each node has more neighbors in its own community than in any other.

9. Stochastic block models generate networks with group structure. Communities in a network can be recovered by fitting a stochastic block model to it. This is done by maximizing the likelihood that the network is reproduced by the model. This approach needs to have the number of groups as input, but there are suitable procedures to estimate it.

10. Community detection algorithms can be evaluated by testing if they can recover the known community structure of benchmark networks. The popular GN and LFR artificial benchmark networks are derived from special stochastic block models. Real networks with group attributes may or may not be useful for testing, depending on whether the attributes of the nodes are a factor in the genesis of the community structure. Partition similarity measures are used to evaluate how closely the communities detected by an algorithm resemble those in a benchmark.

6.6 Further Reading

For a thorough exposition of the topic, we refer to a number of review articles (Porter et al., 2009; Fortunato, 2010; Fortunato and Hric, 2016). A comprehensive introduction to definitions of communities in social network analysis can be found in Wasserman and Faust (1994). We also recommend reading Zachary's paper on the analysis of the karate club network (Zachary, 1977).

Strong communities were introduced by Luccio and Sami (1969). Radicchi et al. (2004) relaxed the concept of strong community by defining weak communities. Hu et al. (2008) proposed less stringent notions for strong and weak communities.

For network partitioning we refer to the book by Bichot and Siarry (2013). The Kernighan–Lin algorithm was originally proposed by Kernighan and Lin (1970). Jain et al. (1999) and Xu and Wunsch (2008) provide good introductions to data clustering.

The Girvan–Newman algorithm was introduced by Girvan and Newman (2002). Modularity was defined by Newman and Girvan (2004) and the greedy method for modularity optimization was presented by Newman (2004a). A fast version of Newman's greedy technique was proposed by Clauset *et al.* (2004). The Louvain method was developed by Blondel *et al.* (2008).

Guimerà *et al.* (2004) discovered that partitions of random graphs can attain large modularity scores. The resolution limit of modularity maximization was exposed by Fortunato and Barthélemy (2007). The extension of modularity to the case of weighted networks was proposed by Newman (2004b), while Arenas *et al.* (2007) further extended the definition to networks that are also directed.

Multiresolution modularity optimization was initiated by Reichardt and Bornholdt (2006) and Arenas *et al.* (2008). Raghavan *et al.* (2007) developed the label propagation method. Stochastic block models were introduced in foundational papers by Fienberg and Wasserman (1981) and Holland *et al.* (1983). Modern methods to fit stochastic block models to networks and infer the communities were presented by Karrer and Newman (2011) and Peixoto (2014).

The planted partition model was an idea of Condon and Karp (2001), and the GN benchmark based on it is named after Girvan and Newman (2002). The LFR benchmark was developed by Lancichinetti *et al.* (2008) and is named after the three authors. Lancichinetti and Fortunato (2009) carried out a comparative analysis of many algorithms on the LFR benchmark. Yang and Leskovec (2012) and Hric *et al.* (2014) found a disconnect between structural communities and attribute-based communities. Meilă (2007) provides a good review of partition similarity measures. The fraction of correctly detected nodes and the normalized mutual information were introduced by Girvan and Newman (2002) and Fred and Jain (2003), respectively.

Exercises

6.1 Go through the Chapter 6 Tutorial on the book's GitHub repository.[3]

6.2 Strong communities are also weak communities but the converse is not true, in general. Provide an example of a weak community which is not a strong community.

6.3 The definition of a weak community captures our naive expectation that there must be more links inside than outside a community. However, for a subnetwork C to be a weak community it is not necessary that the number of internal links L_C exceeds that of external links k_C^{ext}. What is the actual condition? Provide a small example of a weak community C such that $L_C < k_C^{ext}$.

6.4 Any partition of a network with N nodes into $N - 1$ groups must necessarily consist of one pair of nodes and singletons. How many such partitions are there?

6.5 Suppose that a network is a giant clique of N nodes, with N even. What is the solution of the graph bisection problem for this network? What is the cut size of the resulting bipartition?

---

[3] github.com/CambridgeUniversityPress/FirstCourseNetworkScience

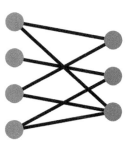

Fig. 6.18   Schematic example of a bipartite graph.

6.6   In graph bisection, minimizing the cut size between the two clusters implies maximizing the number of links inside the clusters. So, network partitioning may appear to be equivalent to community detection when we deal with bipartitions. Explain why this is not true and provide an example.

6.7   Find the best bisection of Zachary's karate club network by applying the Kernighan–Lin algorithm. You can use the `kernighan_lin_bisection()` function of NetworkX. Compare the resulting bipartition with the natural partition of the network and expose similarities and differences.

6.8   In every dendrogram produced by agglomerative hierarchical clustering, a horizontal line indicates the merger of two groups of nodes. If the network has N nodes, what is the total number of possible horizontal lines of the dendrogram?

6.9   Compare two community detection methods on Zachary's karate club network. First, apply the Girvan–Newman algorithm using the `community.girvan_newman()` function of NetworkX. Verify that the partition P_{NG} in five clusters is the one with highest modularity. (*Hint*: See the tutorial for this chapter.) Second, apply the NetworkX function for greedy modularity optimization, `community.greedy_modularity_communities()`. How many communities are there in the resulting partition P_G? Which partition has higher modularity, P_{NG} or P_G?

6.10   Recall that a bipartite network consists of two classes of nodes, say A and B, and links join only nodes in A with nodes in B. See the example in Figure 6.18, where the classes are colored in red and blue. Show that the modularity of the partition of a bipartite network in the two groups A and B is $-1/2$. This is also the lowest value that modularity can reach.

6.11   Suppose you have a clique with N nodes. Show that any bipartition of the clique has negative modularity. For extra credit, show that modularity is negative for *any* partition of the clique into more than one cluster. (*Hint*: Let the two groups have N_A and $N_B = N - N_A$ nodes, respectively. The result holds for any value of N_A.) Since the partition in a single cluster has modularity zero, it is also the maximum modularity partition: *modularity optimization does not split cliques!*

6.12 Calculate the modularity for the two partitions of the ring of cliques in Figure 6.13: the one in which each clique is a community and the one in which the cliques are paired. Confirm the scores reported in the caption.

6.13 Suppose that A and B are two of the $q > 2$ clusters of a network partition, with degrees k_A and k_B, respectively. For simplicity, assume that k_A and k_B are approximately the same: $k_A \approx k_B = k^*$. Let L_A^{int}, L_B^{int}, L_{AB} be the number of links inside cluster A, inside cluster B, and between A and B, respectively. Compute the difference in modularity between this partition and the one in which A and B are merged. (*Hint*: Since modularity is a sum over the clusters, you can neglect the contributions coming from all clusters other than A and B, which are the same for both partitions and cancel out in the difference.) What condition on k^* makes the partition with A and B merged have larger modularity than the one in which they are separated? The resolution limit of modularity results from this condition and applies to pairs of clusters that are connected by at least one link ($L_{AB} > 0$).

6.14 Apply the label propagation algorithm on Zachary's karate club network. Use the `asyn_lpa_communities()` function of NetworkX. Compare the result with the natural partition of the network.

6.15 Suppose that the $q \times q$ matrix of a stochastic block model has non-zero entries only on the diagonal. What can we conclude about the networks generated by the model?

6.16 Write a program that, given a value of the expected external degree $\langle k^{ext} \rangle$ in input, generates a GN benchmark. Follow these steps:
1. Assign labels to nodes in the same group. So, nodes 0 through 31 have label c_1, nodes 32 through 63 have label c_2, nodes 64 through 95 have label c_3, and nodes 96 through 127 have label c_4.
2. Calculate the probabilities $p_{ext} = \langle k^{ext} \rangle / 96$ and $p_{int} = (16 - \langle k^{ext} \rangle)/31$.
3. Loop over all pairs of nodes. If two nodes have the same label (i.e. they are in the same group) add a link with probability p_{int}, otherwise with probability p_{ext}.

6.17 Test the Louvain algorithm on the GN benchmark. (*Hint:* You can install and import the `community` module from the `python-louvain` package as illustrated in Section 6.3.2.) To construct the benchmark, use the procedure described in Exercise 6.16. Use the following values for the expected external degree of the GN benchmark: $\langle k^{ext} \rangle = 2, 4, 6, 8, 10, 12, 14$. For each value, generate 10 different configurations of the benchmark and apply the Louvain algorithm on each of them. Use the fraction of correctly detected nodes (defined in Section 6.4) to compute the similarity of each partition to the planted partition of the benchmark. Calculate the average and the standard deviation of the similarity over the 10 realizations and plot the average as a function of $\langle k^{ext} \rangle$, using the standard deviation for error bars. What do you observe? Why?

6.18 Create Erdős–Rényi random networks with $N = 1000$ nodes and $L = 5000$, 10,000, 15,000, 20,000, 25,000, and 30,000 links. Apply the Louvain algorithm to each network. (*Hint*: You can install and import the `community` module from the `python-louvain` package as illustrated in Section 6.3.2.) Check the modularity values of the resulting partitions: are they close to zero? Plot the number of communities as a function of the average degree of the random networks: what conclusion does the trend suggest?

6.19 Consider a network with N nodes. Compute the similarity between the partition in which the network is a single community and the partition of singletons, using the normalized mutual information. Does this depend on the structure of the network; that is, does a random network give a different result from a complete network?

6.20 Sometimes a network is too large to perform a quick community analysis, and you may wish to work on a subnetwork based on a sample of the nodes and all links among them. There are multiple ways to sample nodes. A *random sample* is obtained by considering each node with the same probability, irrespective of the network structure. A *snowball sample* is obtained by starting from one or a few nodes, then adding their neighbors, and so on until enough nodes have been included. This can be done using the breadth-first search algorithm discussed in Section 2.5. Take one of the large datasets available on the book's GitHub repository, for example the IMDB co-star network, and construct two subnetworks with $N = 1000$ nodes using these two sampling methods.
   1. Compare the densities of the two subnetworks: are they the same? Why or why not?
   2. Compare the average path lengths of the two subnetworks: are they the same? Why or why not?
   3. Compare the degree distributions of the two subnetworks: are they the same? Why or why not?

6.21 Use the IMDB co-star dataset available on the book's GitHub repository to perform a community analysis of the network shown in Figure 0.2(b). What do the major communities correspond to, in terms of features shared among the actors and actresses — genres? periods? languages or countries of origin? What algorithms work best? Do they yield consistent results? (*Hint 1*: The network is large, so you may want to start with a subnetwork based on a sample of nodes, as described in Exercise 6.20. *Hint 2*: You can find IDs of movie stars in the data files, then search on imdb.com to get more information. For example, Marilyn Monroe has ID nm0000054 and her details can be found at imdb.com/name/nm0000054.)

6.22 Analyze the community structure of the Enron email network (the dataset is available on the book's GitHub repository, in the email-Enron folder). Can you identify the module at the right-hand side in Figure 0.4? (*Hint*: Treat the network as undirected. It is large, so you should focus on the core as shown in Figure 0.4; use the NetworkX function `k_core()` from Section 3.6 with $k = 43$. You will first need to remove self-loops.)

6.23 Analyze the community structure of the Wikipedia math network (the dataset is available on the book's GitHub repository, in the enwiki_math folder). Discuss the topics of the different communities you observe in Figure 0.5. (*Hint*: Treat the network as undirected. Since it is large, follow the hint in Exercise 6.22 to focus on the core with $k = 30$. You can find article titles using the node attribute "title" after reading the enwiki_math.graphml file.)

6.24 Analyze the community structure of the Internet router network (the dataset is available on the book's GitHub repository, in the tech-RL-caida folder). The Louvain algorithm was used to detect communities identified by the different colors in Figure 0.6. Investigate the community with the largest hubs. What is its average degree? (*Hint*: This network is very large, so you may wish to run core decomposition first, as described in the previous exercises. Find a suitable value of k so that a couple of thousand nodes remain in the network core.)

## 7

Dynamics

dyn·am·ics: (*n.*) the forces or properties that stimulate growth, development, or change within a system or process.

Four days before the 2016 US election, a conspiracy site posted a false news report alleging that the staff of one of the presidential candidates was engaged in satanic rituals. The fake news spread in viral fashion on Twitter, mainly among supporters of the opposing candidate who accepted as fact fabricated stories that reinforced their beliefs. Automated accounts known as social bots also contributed to the spread by amplifying its reach. This was just one example of thousands of false news items circulating during the electoral campaign — a real epidemic that influenced opinions and, according to some experts, might even have affected the election results.

Scientists are studying the factors that make people and social media platforms vulnerable to this kind of manipulation. Network scientists, in particular, study these phenomena because the structure of online social networks plays a key role in the viral nature of certain messages. For example, Figure 7.1 shows a portion of the diffusion network for the fake report mentioned above. We immediately notice that some nodes, including social bots, were particularly influential.

The spread of misinformation is a special case of information diffusion, one class of dynamic processes that take place on networks. This chapter considers a few other important types of network processes in addition to information diffusion: epidemics, opinion formation, and search. In each case we focus on the *dynamics*, that is, what happens on the network over time — how information and diseases are transmitted across links, how node attributes are affected by their interactions, and how one can search or navigate networks. We present several *models* that capture these dynamics.

7.1 Ideas, Information, Influence

Networks play a central role in the way ideas and information spread in a social community. We are often exposed to new things via friends: for instance, we can find out about a new smartphone model because our best friend just bought it, or discover the latest news on US foreign policy because a friend tells us about it or forwards an article she just read.

Indeed, a lot of what we do is directly or indirectly determined by our social contacts. Social influence is a critical factor when we adopt a behavior, make a decision, embrace an innovation, or shape our cultural, political, and religious views. Therefore, modeling

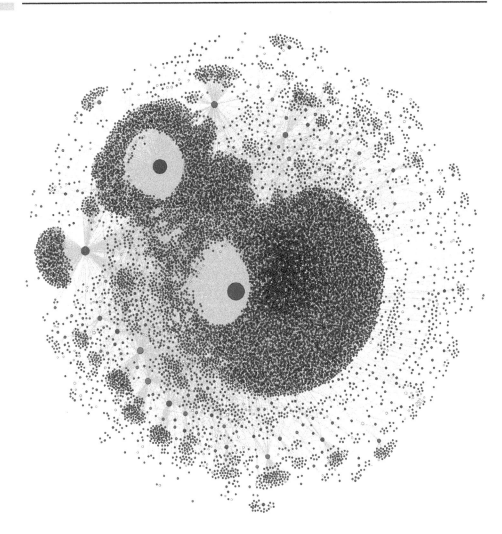

Fig. 7.1 Core of the diffusion network of a viral fabricated news report entitled *"Spirit cooking": Clinton campaign chairman practices bizarre occult ritual,* published by the conspiracy site InfoWars four days before the 2016 US election and shared in over 30,000 tweets. Nodes represent Twitter accounts. A link between two nodes indicates that one of the corresponding accounts has retweeted a post by the other containing the article. Node size indicates account influence, measured by the number of times an account sharing the article was retweeted (out-strength). Node color represents the likelihood that an account is automated, from blue (likely human) to red (likely bot); yellow nodes could not be evaluated because they had been suspended. Recall from Chapter 4 that Twitter does not provide data to reconstruct a retweet tree; all retweets point to the original tweet. The retweet network shown here combines multiple cascades (each a star network originating from a different tweet) that all share the same article. Image courtesy of Shao *et al.* (2018b); an interactive version of this network is available online (iunetsci.github.io/HoaxyBots/).

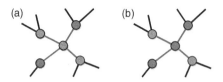

Diffusion of social influence on networks. (a) The central (gray) node is inactive and has two active (red) and two inactive (gray) neighbors. (b) The node becomes active due to the influence of its active neighbors.

how influence, ideas, and information spread in social networks is a key application of network science. These spreading processes are also called *social contagion*, because they resemble a disease that is transmitted via contacts between individuals. In fact, as we shall see in Section 7.2.2, social contagion is often modeled as the spreading of an epidemic.

In any model of influence spreading, we assume that a certain number of nodes (*influencers*) are initially activated, representing that they adopted a new idea, innovation, behavior, etc. Then each inactive node is activated (or not) according to some rule that depends on the presence of active neighbors and on other conditions and parameters, as illustrated in Figure 7.2. The outcome of this process is the generation of *influence cascades*, the activation in sequence of a subset of the nodes in the network. Cascades can range from a handful of nodes to *global cascades* involving a substantial proportion of the network. Sometimes a few nodes end up influencing the whole network. In Section 4.5 we discussed the structure of cascade networks; to see *how* these cascades unfold over time, let us discuss two main classes of social contagion models based on *thresholds* and *independent cascades*.

7.1.1 Threshold Models

The principle of threshold models is very simple: a node can be activated only if the influence exerted on it by its active neighbors exceeds a value. In the most basic version, the *linear threshold model*, the influence on a node is defined as a sum over its active neighbors, in which the contribution of each neighbor is given by the weight of the link joining it to the node: the stronger the connection, the higher the influence of the neighbor. If the influence surpasses a node-specific threshold, the node becomes active, meaning that it adopts the idea, information, or behavior.

In the linear threshold model, the influence on a node i is expressed by

$$I(i) = \sum_{j:\text{active}} w_{ji}. \tag{7.1}$$

In Eq. (7.1) the sum includes only active neighbors of i; if a node j is not a neighbor, there is no link joining it to i and $w_{ji} = 0$. The condition for the activation of i is

$$I(i) \geq \theta_i, \tag{7.2}$$

where θ_i is the specific threshold of node i, which is assigned to the node before the process starts. Such a threshold indicates the tendency of an individual to be influenced, which usually varies from one individual to another. If the graph is unweighted, Eq. (7.2) reduces to

$$n_i^{on} \geq \theta_i, \qquad\qquad (7.3)$$

where n_i^{on} is the number of active neighbors of i. In this case, if the number of active neighbors is above the node's threshold, the node is activated, otherwise it remains inactive. If all nodes have equal threshold θ, Eq. (7.3) turns into the simple condition that any inactive node must have at least θ active neighbors to become active.

The model works as follows. First, we choose our network, which could originate from real data or from a graph generation model like the ones introduced in Chapter 5. For simplicity, let us assume that the graph is not weighted. Next we assign a threshold to all nodes, for instance by generating random numbers in some interval. Then a given number of nodes is activated; again, they can be selected at random. Finally, we go through iterative steps in which inactive nodes can become active based on the activation of their neighbors.

Each iteration of the model dynamics consists of the following operations:

1. All active nodes remain active.
2. Each inactive node is activated if the number of active neighbors is at or above its threshold.

The steps are repeated until no further nodes can be activated.

The order in which nodes are considered should not affect the outcome in models of network dynamics. There are two ways to ensure this when implementing the node update rules. In *asynchronous* implementations, nodes are evaluated in a different random sequence at each iteration. This is to avoid biases that may result from always following the same sequence. In *synchronous* implementations, the new activation state of each node in each iteration is determined using the activation values of the other nodes from the previous iteration; all of the nodes are then updated at the end of the iteration. The order is irrelevant in this case.

Many variations of the linear threshold model have been proposed. In the *fractional threshold* model, we consider the fraction rather than the number of active neighbors. So, in this model, in order to activate a node with threshold $1/2$, say, at least half of its neighbors must be active. Figure 7.3 shows how the dynamics of the model unfold on a simple network: the activation of one node triggers a cascade that eventually leads to the activation of all other nodes.

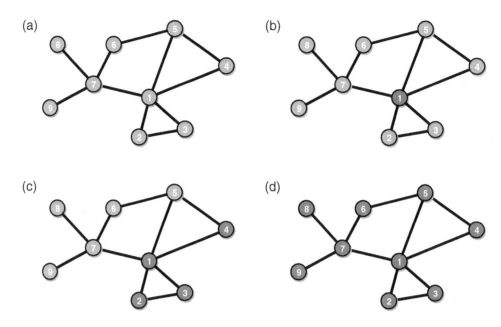

Fig. 7.3 Fractional threshold model of influence diffusion. The activation threshold is $1/2$ for all nodes. (a) Initially, all nodes are inactive. (b) Node **1** is activated. (c) Nodes **2**, **3**, and **4** have two neighbors and one of them is **1**, which is active, so they get activated. (d) After the activation of **4**, node **5** has two active neighbors out of three and becomes active (since $2/3 \geq 1/2$). Likewise, nodes **6**, **7**, **8**, and **9** are activated subsequently.

In the fractional threshold model, the activation condition is

$$\frac{n_i^{on}}{k_i} \geq \theta_i, \tag{7.4}$$

where k_i is the degree of node i. The ratio on the left-hand side of Eq. (7.4) is the fraction of active neighbors of i. If all nodes have equal threshold θ, the condition is that an inactive node needs to have at least a fraction θ of active neighbors to be activated.

If the network is sparse, whether or not a global cascade is triggered depends on its structure. The key drivers are the *vulnerable nodes* (i.e. those which can be activated by a single active neighbor).

From Eq. (7.4) we see that a node is vulnerable if $k_i \leq 1/\theta_i$, that is, if its degree is below or at the inverse of its threshold.

To have global cascades, the number of vulnerable nodes has to be sufficiently large. Hubs are usually very effective influencers: the higher the number of neighbors, the more likely it is that some of them have sufficiently low degree to be vulnerable. However, being a hub is not always a sufficient condition for influence. The position of the influencer in the network is also important: a cascade in the periphery of the network will hardly manage to work its way through the core.

Another aspect of the network structure that plays an important role in the size of a cascade is the density and separation between communities. The spread is facilitated within dense communities, but hindered across communities. Cluster boundaries act like walls, because a node is unlikely to have multiple active neighbors in different communities.

Knowing the structure of the network enables us to control the size of cascades. In the example of Figure 7.3, if the initial influencer is node **7**, its neighbors **6**, **8**, and **9** will become active, but the cascade stops there, because the fractions of active neighbors of nodes **1** and **5** are $1/5$ and $1/3$, respectively, both below $1/2$. However, if we also manage to successively activate node **2**, say, node **3** would also become active and nodes **2**, **3**, and **7** would activate node **1**, allowing the cascade to propagate to the whole network. So, in this case, influencing node **2** "unblocks" the cascade. Indeed, the success of a product or idea often depends on the identification of key individuals that need to be persuaded to buy it. This issue is central in viral marketing, where social networks are used to promote products. Appendix B.6 presents a demonstration of the fractional threshold model.

7.1.2 Independent Cascade Models

Threshold models are based on the concept of *peer pressure*: the more of our contacts share an idea or own a product, the more likely it is that we will adopt it ourselves. It is as if our active social neighbors work together to persuade us. But social influence is often one-to-one: we may be convinced to adopt a product or belief if a single friend speaks enthusiastically about it. Each of our other contacts will have their own influence, unless we have already bought the product or the idea. *Independent cascade models* focus on such node–node interactions.

The setup is the same as for threshold models, in that a network is chosen, or built, and some of the nodes are activated. As soon as a node becomes active, it has one chance to "convince" each of its inactive neighbors; each neighbor is activated with some *influence probability*. If a node fails to activate its friend, it cannot try again. However, the friend can still be persuaded by another active neighbor. The process, illustrated in Figure 7.4, goes on until no further activations occur.

In the simplest version of independent cascade models, an active node i has a probability p_{ij} of convincing its inactive neighbor j. Such probability generally depends only on the specific influencer–neighbor pair, so the outcome of each interaction is not affected by what happens to the other pairs. In asynchronous implementations, if j has multiple active neighbors, their activation attempts are sequenced in an arbitrary order to avoid bias. The influence probabilities p_{ij} and p_{ji} may differ, because each node has its own ability to persuade and susceptibility to be persuaded, in general. So it may be easier for i to influence j than vice versa. The probability p_{ij} can be interpreted as the weight of the link from i to j.

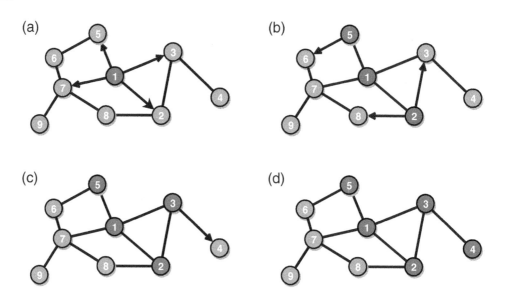

Fig. 7.4 Independent cascade model. The influence probability is set to $1/2$ for all pairs of nodes, so the success of every interaction is decided by flipping a fair coin. The arrows indicate who is trying to influence whom. (a) Node **1** is activated and tries to influence its inactive neighbors **2**, **3**, **5**, and **7**. (b) Nodes **2** and **5** become active and exert their influence on **3**, **6**, and **8**. (c) Node **3** is activated and attempts to convince **4**. (d) Node **4** turns active, and the cascade stops.

   Clearly, the higher the number of active neighbors of an inactive target node, the larger the number of attempts to influence the node and the more likely that it will get activated. Consequently, threshold models and independent cascade models are related, but there are important differences. Threshold models are centered on the target, who is activated if the threshold condition is satisfied. Independent cascade models are centered on the influencer, who persuades its inactive neighbors with given probabilities. In addition, threshold models are usually *deterministic*. The activation of any node depends on whether the threshold condition is satisfied or not; chance plays no role. This means that, if we start from the same initial set of active nodes and activate nodes synchronously, there can only be one outcome. Independent cascade models are instead *probabilistic*: the unfolding of the dynamics depends on chance. In the example of Figure 7.4, different cascades could be triggered by the initial activation of node **1**. In an independent cascade model we can "unblock" a cascade by activating further nodes, suitably chosen, as we have seen in Section 7.1.1 for the linear threshold model. However, due to the model's probabilistic character, it is hard to make predictions about the future progress of the cascade, even when the network structure is known.

   The very simple models we have described cannot be expected to reproduce real social contagion dynamics. However, more sophisticated variations of these models are capable of capturing important features of many real-world phenomena. One example is a probabilistic version of the threshold model, in which the chances of activation grow with the number of active neighbors. This is similar to the independent cascade model,

but contacts with active neighbors are not independent of each other. Such a mechanism models so-called *complex contagion* processes: each new person that exposes us
to a product or idea has greater influence than previous ones in getting us to adopt it or
believe it.

7.2 Epidemic Spreading

In the middle of the fourteenth century, humankind suffered one of the greatest calami-
ties in history: the Black Death. Also known as the Great Plague, it is believed to have
been caused by the bacterium *Yersinia pestis*, carried by fleas living on black rats that
were regularly traveling on board merchant ships. It probably started in Central Asia and
spread throughout all of Europe between 1346 and 1353 (Figure 7.5). The Black Death is
estimated to have killed 30–60% of Europe's population.

While the potentially devastating effects of infectious diseases have been effectively
mitigated by great improvements in human living conditions and progress in medicine
and biology, especially over the past century, the speed of their spread has been strongly
enhanced by technological advances in human transportation. In the Middle Ages, the most
effective means of travel were horses on land and ships on the sea, and it would take months
to reach a remote destination. Nowadays, it takes just a few hours to fly across continents.
A person contracting Ebola in Africa could easily travel to Europe, Asia, or America and
spread the disease there while still being unaware of it. The world has been facing this kind
of emergency repeatedly in recent years.

Technology has also created new forms of epidemics. Computer viruses and other mal-
ware spread through the Internet, compromising the function of millions of devices. Mobile
phone viruses can easily be transmitted via Bluetooth or Multimedia Messaging Services
(MMS). Online social media have become fertile ground for the diffusion of rumors,
hoaxes, fake news, conspiracies, and junk science. Information-spreading processes bear
many similarities with the epidemics of infectious diseases.

Epidemics spread on *contact networks*, such as networks of physical contacts
(Figure 7.6), transportation (Figure 0.7), the Internet (Figure 0.6), email (Figure 0.4), online
social networks (Figures 0.1 and 0.3), and mobile phone communication. Many such networks are characterized by the presence of hubs (discussed in Chapter 3), which play a
central role in the process. In the remainder of this section we review classic models of
epidemic spreading and point out the key differences in the dynamics when they unfold in
networks.

7.2.1 SIS and SIR Models

Classic epidemic models divide the population into different compartments, corresponding to different stages of the disease. The two key compartments are *susceptible* (S) and
infected (I). Susceptible individuals can contract the disease, infected individuals have

1346   1347   1348   1349   1350   1351   1352   1353

~ ~ ~ ~ Approximate border between the Principality
of Kiev and the Golden Horde - passage
prohibited for Christians.

Land trade routes

Maritime trade routes

Fig. 7.5 The Black Death reached Europe in 1346 and spread over the whole continent within a few years. The map shows the regions hit by the disease over time, as well as the likely routes of its migration. Image by Flappiefh licensed under CC-BY-SA 4.0 (commons.wikimedia.org/wiki/File:1346-1353_spread_of_the_Black_Death_in_Europe_map.svg).

contracted it already and can transmit it to susceptible individuals. Depending on what kind of disease we consider, additional compartments may be needed. In the *susceptible–infected–susceptible (SIS) model*, infected individuals become susceptible again when they recover from the disease, so they can contract it again (Figure 7.7). The model applies to diseases that do not confer long-lasting immunity, like the common cold.

The SIS model starts with either a real-world contact network, reconstructed from empirical data, or an artificial network generated by some model, like those presented in Chapter 5. Next, we assume that some of the nodes are infected according to some criterion (e.g. at random). All the other nodes are susceptible. During the model dynamics, susceptible individuals contract the disease with a certain probability called the *infection rate* at each encounter with an infected individual. Infected people recover from the disease, turning to susceptible, with some probability called *recovery rate* at each time step.

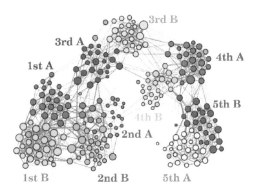

Fig. 7.6 Contact network in a primary school. The links indicate face-to-face proximity among children and teachers in a French school, tracked by radio-frequency identification devices. The colors label children in the same class and grade; teachers are shown in gray. Nodes with higher degree have larger size, contacts of longer duration are represented by thicker links. While every child eventually interacts with all their classmates after a sufficiently long time, some of them engage with children of other classes as well. This type of network may suggest interventions aimed at containing or mitigating the propagation of infectious diseases in schools. Image reprinted from Stehlé *et al.* (2011) under CC-BY-4.0 license.

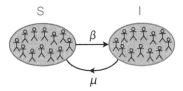

Fig. 7.7 Compartments and transitions in the SIS model. Each susceptible individual gets the disease with probability β after each contact with an infected individual. At each time step, each infected individual has a probability μ of recovering from the disease and becoming susceptible again. Individuals can be infected multiple times.

In each iteration of the SIS dynamics, we visit all nodes. For each node i:

1. If i is susceptible, loop over its neighbors. For each infected neighbor, i becomes infected with probability β.
2. If i is infected, i becomes susceptible with probability μ.

As in other spreading models, nodes can be visited asynchronously in random order, or synchronously. The *infection rate* β and the *recovery rate* μ are the key parameters of the model.

The dynamics produce a number of transitions from S to I and from I to S that can be sustained indefinitely, under suitable conditions.

Another classic model is the *susceptible–infected–recovered (SIR) model*. When infected individuals recover from the disease, they move to a third compartment of *recovered* (R) people, and cannot be infected anymore (Figure 7.8). The model applies to diseases that confer long-lasting immunity, like measles, mumps, rubella, and so on. Note

Fig. 7.8 Compartments and transitions in the SIR model. Each susceptible individual gets the disease with probability β after each contact with an infected individual. Each infected individual has probability μ of recovering (or dying) from the disease at each time step.

that death is a special case of the recovered state for deadly diseases, because deceased people do not infect others. The dynamics of infection and recovery follow closely that of the SIS model above, with the same infection rate and recovery rate parameters. The only difference is that, when an infected individual recovers, it is moved to the R state rather than back to the I state; it will not play any further role in the dynamics. Eventually, the SIR model spreading stops, when there are no more infected individuals.

In Figure 7.9 we see the characteristic evolution of both SIS and SIR models, plotted by the fraction of the population that has contracted the disease as a function of time. Initially, just a few people are infected, and the diffusion of the epidemic is irregular and slow. This is followed by a ramp-up phase of exponential growth, which can quickly affect a large number of people. Finally, the process reaches a stationary state, in which the disease is either *endemic* (i.e. it affects a stable fraction of the population over time) or eradicated.

Classic epidemiology models can be simplified by making the *homogeneous mixing approximation*, which consists of assuming that each individual can be in contact with any other. This way, all individuals in the same compartment have identical behavior and only the relative proportions of people in the various compartments matter for the model dynamics. This is equivalent to assuming that the individuals are nodes of a complete graph, where everybody is linked to everyone else. Such a simplifying assumption could be justified for a small population, like the inhabitants of a little village where all people are in touch with each other. But in real, large-scale epidemics, individuals can only be infected by the people they come in contact with. It is therefore critical to reconstruct the actual network of contacts to the extent possible.

At each iteration of the model, there are newly infected individuals, called *secondary infections*, along with sick individuals who recover from the disease. For the epidemic to spread, there must be more secondary infections than recovered people, because only in this way can the number of infected people grow. On homogeneous networks, where all nodes have similar degree, meaning that every individual comes in contact with roughly the same number of people, this condition leads to a *threshold effect*. We can define the *basic reproduction number* as the average number of newly infected people generated by an infected individual over the course of its infectious period. This quantity depends on the infection rate, the recovery rate, and the average degree. If it is larger than a threshold, then the epidemic can hit a significant fraction of the population; otherwise it will be absorbed quickly, without major effects.

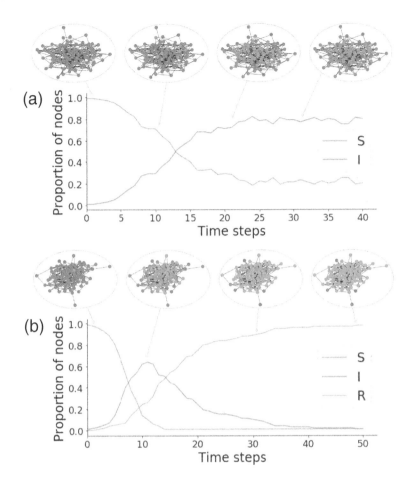

Fig. 7.9 Schematic evolution of (a) SIS and (b) SIR model dynamics. The fraction of infected individuals is plotted versus time, following an epidemic outbreak. After an initial phase, characterized by a low proportion of infected people, the epidemic grows fast until a fraction of the population is hit by the disease. The final phase depends on the model: for the SIS model, the infected stabilize around a constant fraction (which can also be very small or even zero), signaling an endemic state. For the SIR model, the infected fraction always goes down to zero as individuals recover.

Assume a homogeneous contact network, with all nodes having degree approximately equal to the average $\langle k \rangle$. According to the SIS and the SIR model dynamics, each sick person infects a susceptible neighbor with probability β. In the early stages of the epidemic only a few people are infected, so we can assume that each of them is in contact with mostly susceptible individuals. Each infected person can transmit the disease to about $\langle k \rangle$ individuals at each iteration. Therefore, the average number of infections caused by a single person after one iteration in the early stage of the spreading process is $\beta \langle k \rangle$. In contrast, at each iteration every sick individual recovers with probability μ. So, if there are I infected individuals, after one iteration there will be on average

$I_{sec} = \beta \langle k \rangle I$ secondary infections, while $I_{rec} = \mu I$ people are expected to recover. For the epidemic to spread we must have $I_{sec} > I_{rec}$, which leads to the *epidemic threshold* condition:

$$\beta \langle k \rangle I > \mu I \implies R_0 = \frac{\beta}{\mu} \langle k \rangle > 1. \tag{7.5}$$

The variable $R_0 = \beta \langle k \rangle / \mu$ is the *basic reproduction number*. Equation (7.5) states that if $R_0 < 1$, the initial outbreak dies out in a short time, affecting only a few individuals. If $R_0 > 1$, the epidemic can keep spreading.

For the epidemic to affect a significant portion of the population, each infected person must transmit the disease to more than one other individual. This condition is necessary but not sufficient: in certain situations, the epidemic may not have major consequences, even if the basic reproduction number is above one; factors such as quarantine policies or the network community structure might prevent the epidemic from spreading. In general, the higher the basic reproduction number, the more infectious the disease. For example, the number is above 10 for measles and around two for Ebola.

Appendix B.5 presents a demonstration of both SIS and SIR models on homogeneous networks. But as we have seen, real contact networks are not homogeneous. The presence of hubs changes the scenario significantly. If there are nodes with very large degree, *there is effectively no threshold*: even diseases with low infection rate and/or high recovery rate may end up affecting a sizable fraction of the population! In fact, even if the probability of contracting the disease is low, it is fairly easy to infect one or more hubs, who are very exposed due to their high number of contacts. Once infected, the hubs are dangerous spreaders among their many susceptible contacts, who will propagate the infection further to their contacts, and so on.

Because of the role of hubs, when facing real epidemic emergencies, effective containment strategies should aim to vaccinate or isolate people with many contacts. For example, sex workers are primary targets of vaccination campaigns for sexually transmitted infections. In many cases, it is not obvious how to identify contact network hubs. Section 3.3 suggests a way. By following the links of a network we increase the chance of bumping into hubs. So, instead of vaccinating a random sample of the population, one should vaccinate their friends!

7.2.2 Rumor Spreading

Social contagion can naturally be described as the spreading of an epidemic. In fact, the models of social contagion we have reviewed share similarities with the SIS and the SIR models, especially the independent cascade models of Section 7.1.2.

A variant of the SIR model can be used to describe the spreading of rumors in a community. Like the SIR model, this *rumor-spreading model* has three compartments: ignorant (S), spreaders (I), and stiflers (R). The latter are people who know about the rumor but do not contribute to spreading it. The basic idea is that people are engaged in the diffusion of

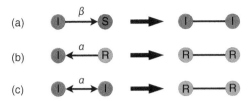

Fig. 7.10 Rumor-spreading model. (a) The rumor circulates only if a spreader (I) meets an ignorant (S). In this case, the ignorant becomes a spreader with probability β. (b) If a spreader meets a stifler (R), the spreader becomes a stifler with probability α. (c) If two spreaders meet they both become stiflers with probability α.

the rumor as long as they find people who are unaware of it, otherwise they lose interest and stop spreading the rumor.

The rumor-spreading model starts with a network of contacts, which could be a real network or an artificial one generated by some model, like those we have seen in Chapter 5. All the nodes are ignorant except for some that are turned into spreaders of the rumor, according to some criterion; they can be selected at random. In the dynamics of the model, when a spreader approaches an ignorant, the rumor is told and the ignorant becomes a spreader with a *transmission probability*. When a spreader meets a stifler, the spreader becomes a stifler with a *stop probability*. When two spreaders meet, they both turn to stiflers with the same stop probability. Figure 7.10 illustrates these transitions. Nothing happens if an ignorant meets a stifler.

In each iteration of the rumor-spreading model dynamics, all nodes are visited synchronously or asynchronously in random order. For each node i:

1. If i is ignorant, loop over its neighbors. For each spreader neighbor, i becomes a spreader with probability β.
2. If i is a spreader, loop over its neighbors.

   (i) For each stifler neighbor, i becomes a stifler with probability α.
   (ii) For each spreader neighbor, i and the neighbor both become stiflers with probability α.

The transmission probability β and stop probability α are the two key parameters of the model.

An important difference from the SIR model is that here the transition from I to R does not occur spontaneously (in that a sick person recovers from the disease), but depends on the interaction between individuals. As in the SIR model, starting from a few spreaders, eventually all individuals will be either ignorant or stiflers, as in this case the dynamics cannot produce any change. The number of stiflers in the final state is also the number of people who found out about the rumor.

The rumor-spreading model does not have a threshold effect, even on homogeneous networks. The rumor can reach a large number of people, even if the transmission probability is low. On heterogeneous networks, there is still no threshold, and the final number of people aware of the rumor is lower than on homogeneous networks with equal numbers of nodes and links. This occurs because the rumor reaches the hubs in the early stages of the process and they quickly become stiflers due to their multiple interactions with other individuals, some of whom may be aware of the rumor. Once the hubs turn into stiflers, the diffusion process slows down.

7.3 Opinion Dynamics

We have opinions about everybody and everything. Opinions drive our behavior, affect our choices, influence our plans. Policies implemented by governments worldwide are dictated by opinions about trade, conflicts, immigration, pandemics, the environment, and so on. Opinion dynamics are the processes that determine how opinions form and diffuse in society. With the introduction of the Internet and social media, humankind has endowed itself with incredibly powerful tools to circulate and even manipulate opinions. Opinions spread on networks like those of Facebook friends and Twitter followers. Therefore, network models can help us understand how opinions propagate.

Models of opinion dynamics are similar to models of influence spreading seen in the previous section, but they have some distinctive features. We can represent an opinion as a number or a set of numbers. Models are usually divided into two categories, based on whether they use *discrete* (integer) or *continuous* (real-valued) opinions. Next we introduce simple models in both classes. We will also discuss the interplay between network structure and dynamics, as in several realistic scenarios the structure of the network affects the processes that take place on it, but the dynamics may in turn change the structure as well.

7.3.1 Discrete Opinions

People are sometimes confronted with a limited number of positions on a specific issue, often just two positions: right/left, Android/iPhone, buy/sell, and so on. In such cases, the opinion is represented by an integer attribute or *state* of each node. For simplicity, let us consider just the case of binary opinions.

A model is characterized by the set of rules that determine how the opinion of a node changes due to the opinions of its neighbors. The dynamics usually follow these steps:

1. In the initial configuration, opinions are randomly assigned among the nodes of the network. This means that initially there is about the same number of people holding either opinion (*disagreement*).
2. The opinion update rule is applied over and over to all nodes. An iteration consists of running a loop over all nodes. Typically, nodes are updated asynchronously in random order to facilitate convergence.
3. There are two possible outcomes:

   (i) The system reaches a steady state, where no node changes its opinion anymore. Such a state can be a *consensus*, with all nodes having the same opinion, or a *polarization*, with some nodes holding an opinion and the rest holding the other.

(ii) The system does not reach a stationary state, in that some nodes (or all) keep changing their opinions at each iteration. Still, some features of the opinion configuration,
for example the averages of some variables, may stabilize in the long run.

A few standard variables can be computed and monitored in these models:

- The *average opinion* is the arithmetic average of the opinions across the nodes. If we
start from a random distribution of two opinions, zero and one, the average is around 0.5,
as half of the nodes will have opinion zero and the other half opinion one. The average
opinion usually changes during the dynamics and one can keep track of its value after
each iteration. If the system reaches a stationary state, the average converges to a precise
value. If the stationary state is consensus, it equals either zero or one, depending on
which opinion dominates.
- The *exit probability* estimates how likely the network is to reach consensus to opinion
one, as a function of the fraction of nodes with opinion one in the initial configuration.
As an illustration, suppose that we run the model dynamics 100 times, starting from
100 different random configurations. In each initial configuration we assign opinion one
to every node with probability 0.4, so that approximately 40% of the nodes will have
opinion one. Imagine that all runs lead to consensus, 30 of them to consensus opinion
one. The value of the exit probability for initial probability 0.4 of opinion one is then
$30/100 = 0.3$.

Two simple discrete opinion dynamics models are borrowed from statistical physics: the
majority model and the *voter model*. In the former, the dynamics are based on a majority
rule: each node takes the opinion of the majority of its neighbors, as shown in Figure 7.11.
If the number of neighbors is even and there is an equal number of them with either opinion,
then we flip a coin to decide which opinion will be taken by the node. This is basically
equivalent to the fractional threshold model presented in Section 7.1.1, with a threshold of
$1/2$. The difference is one of interpretation: here we think of two opinions in competition
rather than one idea spreading.

Consensus is the stable state in which all nodes have the same opinion and nothing
can change. But there are other stable states: if a node has the opinion of the majority
of its neighbors, as in Figure 7.11, its opinion will not change. Such a local majority
condition is often reached by all nodes in the network, giving rise to stable configurations in which both opinions coexist. On most networks we have seen in this book, like
all model networks of Chapter 5, majority dynamics never reach consensus; the network
gets trapped in states with opinion coexistence. Consensus is reached only on one- and

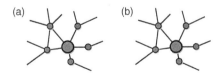

Fig. 7.11 Majority model of opinion dynamics. (a) The node to be updated (big circle) has opinion one (red). The node has five
neighbors: three have opinion one, the other two have opinion zero (blue). (b) The node takes the opinion of the
majority, so it stays red.

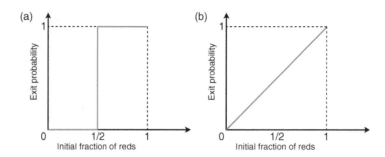

Exit probability. (a) Majority model on a grid network. The step function indicates that the initial proportion of either opinion will determine whether or not the system will reach consensus on that opinion: if the dynamics lead to consensus and more than half of the nodes have opinion one (zero) in the starting configuration, then the consensus is on opinion one (zero). This diagram can be drawn only for one- or two-dimensional grids, as the dynamics never lead to consensus otherwise. (b) Voter model. The diagonal function indicates that the initial proportion of opinions one is also the probability of reaching consensus on opinion one. In contrast with majority dynamics, in the voter model it is possible to reach consensus on any opinion even if less than half of the nodes have that opinion initially.

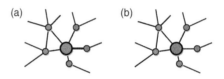

Voter model. The neighborhood of the node to be updated (big circle) is the same as in Figure 7.11. (a) A random neighbor is chosen (the blue node attached to the thick link). (b) The central node takes the opinion of its neighbor.

two-dimensional grids. In fact, on the two-dimensional square grid, consensus is reached in about two-thirds of the runs. If we compute the exit probability for the runs that lead to consensus, we obtain the characteristic step-like profile shown in Figure 7.12(a): in order to reach consensus on any opinion, that opinion must have the majority in the initial configuration.

In the voter model, illustrated in Figure 7.13, each node takes the opinion of a randomly chosen neighbor, whatever it may be. A demonstration of both the majority and voter models is presented in Appendix B.6. Consensus is the only stable state of the voter model dynamics, so it is the inevitable final configuration of the system, on any connected network. In fact, as long as different opinions coexist, neighbors with different opinions can always influence each other. The exit probability of the voter model coincides with the fraction of initial nodes with opinion one, so it is the diagonal function in Figure 7.12(b). In contrast to the majority model, here the outcome of the dynamics is uncertain. For instance, suppose that 30% of the nodes have opinion one in the initial configuration. Then we expect that in 30% of the runs all nodes will end up having opinion one, but we cannot tell in advance whether a specific run will lead to consensus on opinion one or zero.

Many variations of the voter model exist. Common modifications are:

- The presence of *zealots*, nodes who never change their opinion. If they all have the same opinion, they will favor consensus around that opinion, otherwise consensus is never reached.
- Considering more than two opinion states. In this case, the interactions may be constrained to occur only among nodes with sufficiently close opinions. For example one could have three opinions (1, 2, and 3) such that only neighboring opinions can interact (1 and 2, 2 and 3, but not 1 and 3). We discuss such a principle in detail in Section 7.3.2. Non-consensus configurations with non-interacting opinions are stable in any network.
- The possibility for nodes to change their opinion spontaneously, for example with a certain probability at each iteration, on top of the voter dynamics.

Similar modifications can also be applied to other discrete opinion dynamics models.

7.3.2 Continuous Opinions

There are situations in which the opinion of an individual can vary smoothly from one extreme to the other of a range of possible choices. For example, it may express appreciation for an artwork, which could vary continuously from dislike (0) to enthusiasm (10). Or we might wish to model political alignment on a spectrum from very progressive (−1) to very conservative (+1). In such cases, opinions are better represented by real, continuous numbers.

As in discrete opinion models, random opinions are usually assigned to network nodes in the initial configuration. This can be accomplished by generating random numbers in the desired range. Then the opinion values change as they are updated over and over. If at some point the largest variation of any opinion is smaller than a predefined threshold, we can stop the simulation because the system will eventually reach a stationary state. Typical stationary states are *consensus*, *polarization*, or *fragmentation*, depending on whether opinions are concentrated around one, two, or more values, respectively. In the limit of infinite simulation time, each node will have exactly one of the few surviving opinions.

We imagine people having a constructive debate about a topic, with the chance of affecting each other's opinion, especially when their positions are sufficiently close to each other. An individual can hardly convince another if the latter has an opposite point of view. This simple observation has inspired the *principle of bounded confidence*: two opinions can affect each other only if their difference is smaller than a given amount, which is called the *confidence bound*, or *tolerance*.

The original *bounded-confidence model* has an update rule that consists of choosing a node and one of its neighbors. If their opinions differ by less than the confidence bound, they both "move" towards each other, by some relative amount determined by a convergence parameter. Otherwise, the opinions do not change.

In the bounded-confidence model, at iteration t, each node i has opinion $o_i(t)$, which is a real number between, say, zero and one. An iteration consists of a sweep over all nodes, synchronously or in random order. At iteration $t + 1$, when it comes to node i, we pick one of its neighbors at random, say j. If

$$|o_i(t) - o_j(t)| < \epsilon, \tag{7.6}$$

where ϵ is the confidence bound, the values of the opinions are updated to

$$o_i(t + 1) = o_i(t) + \mu[o_j(t) - o_i(t)], \tag{7.7}$$
$$o_j(t + 1) = o_j(t) + \mu[o_i(t) - o_j(t)], \tag{7.8}$$

where $\mu > 0$ is the *convergence parameter*. If $\mu = 1/2$, the opinions converge to their average, meaning that both individuals adopt a common intermediate position. If $\mu = 1$ the opinions switch, in that i adopts j's opinion and vice versa. Usually μ varies in the range $(0, 1/2]$.

If we sum Eqs (7.7) and (7.8) side by side and divide by two, we see that the second terms on the right-hand sides cancel each other out. We conclude that the average opinion of i and j is the same before and after the update: *the average opinion of the system is preserved* in the bounded-confidence dynamics! If the initial opinions are taken at random from the range $[0, 1]$, their average is $1/2$ (with possible small deviations). So, if the system eventually reaches consensus, the opinions of all nodes will cluster around $1/2$.

Starting from a random initial opinion configuration, the dynamics always lead to a stationary state, on any network. The convergence parameter only affects the number of iterations needed to reach convergence. The number of clusters of opinions in the stationary state depends on the confidence bound and on the structure of the network. The lower the confidence bound, the larger the number of final opinion clusters.

For $\epsilon > 1/2$, the system always reaches consensus, on any network, with the opinions centered around $1/2$.

There are many variations of the bounded-confidence model. Common modifications include:

- Using individual values of the confidence bound, to account for the fact that not everybody can be convinced as easily as everybody else. In some extensions, the confidence bound of a node is coupled with the individual's opinion. For instance, if the opinion is close to the extremes of the range, the confidence bound is small because extremists are more difficult to persuade than most people.
- The possibility for individuals to change their opinion spontaneously. As in the voter and other models, this can be implemented by letting nodes change their opinion with some probability at each iteration.

7.3.3 Coevolution of Networks and Dynamics

In Section 2.1 we have seen that assortativity is found in many real graphs, particularly social networks: the nodes are similar to their neighbors. We have also discussed the two possible mechanisms responsible for this: *social influence* (neighbors becoming more similar) and *selection* or *homophily* (similar nodes becoming neighbors). It is plausible that both mechanisms are responsible for the observed assortativity. For instance, if we are constantly debating about an issue with one of our acquaintances, we might either try to find a compromise, or else be better off if we hang around with someone else who shares our view. This happens a lot on social media, where people "unfriend" or "unfollow" contacts with different views. In the models of opinion dynamics discussed so far, the network is fixed. So we are not allowing for selection, because nodes with very similar opinions do not have the option to become neighbors, unless they already are. Similarly, neighbors with very dissimilar opinions cannot become disconnected. Nodes can only influence each other's opinions. A realistic model should allow for the interplay of both influence and selection. This has led to the development of *coevolution models*, in which opinion changes may induce modifications in the network structure, which could in turn affect the opinions, and so on. Basically, opinions and networks *adapt* to each other.

In one coevolution model, opinions are discrete and can take two or more values. At the beginning, opinions are randomly assigned to the nodes. The dynamics consist of alternating selection and influence steps, with a relative frequency determined by a parameter. By selection, nodes set links to other nodes with the same opinion. By influence, nodes take the opinion of their neighbors. Figure 7.14 illustrates the selection and influence steps of the model.

Each iteration of the coevolution model requires a sweep over the nodes, synchronously or in random order. When we examine node i, we select a random neighbor j with different opinion from i:

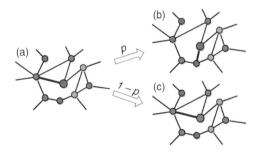

Fig. 7.14   Coevolution of opinions and networks. Opinions are indicated by the colors. (a) A node is chosen (large blue circle in the middle), along with one of its neighbors (red node attached to the thick link). (b) With probability p, the node replaces its neighbor with a node having the same opinion. (c) With probability $1 - p$, the node takes the opinion of the neighbor.

1. With probability p, the link between i and j is rewired from i to a randomly selected non-neighbor holding the same opinion as i (*selection*).
2. Else (with probability $1 - p$), i takes the opinion of j (*influence*).

The *selection probability* p is the single parameter of the model.

Since both selection and influence tend to decrease the number of neighboring node pairs with different opinions, the network eventually reaches a state in which all pairs of neighbors have the same opinion. This means that the network will be divided into a set of separate components, disconnected from one another, with all members of each component holding the same opinion, which may differ across components. We will thus observe a segregation into homogeneous opinion communities, as illustrated in Figure 7.15. Such a scenario is a stable state: no more changes in opinions or network structure take place and the dynamics stop.

When the selection probability is close to zero, influence dominates and the network structure barely changes. The system will basically homogenize the opinions within the connected components of the initial network. When the selection probability is close to one, selection dominates and opinions hardly influence each other. Here the final components of the system are the groups of nodes with the same opinion as in the initial configuration.

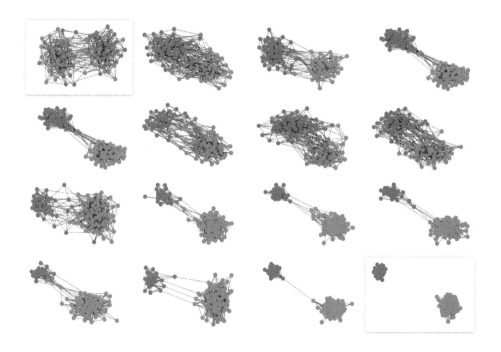

Fig. 7.15 Dynamics of the coevolution model on a network with two communities. Initially (top left), two opinions are randomly distributed among the nodes. The selection probability is $p = 0.7$. Eventually (bottom right), the network becomes segregated into two disconnected components with homogeneous opinions.

Let us see what happens when the number of opinions is large. If we start from a random network with average degree larger than one, we know that it has a giant component (Section 5.1), so for selection probability near zero in the long run there will be a giant community holding the majority opinion, and many small communities with different opinions. For selection probability near one, instead, the link dynamics will break the network into many small components, each made mostly of nodes that were initially assigned one of the distinct opinions. It turns out that there is an abrupt transition between the scenario with a large majority opinion and the scenario with many smaller opinion communities of comparable size. This transition takes place at a threshold value of the selection probability.

Models in which people with similar opinions tend to come together can help us study the emergence of echo chambers in social media, as discussed in Section 4.5 and illustrated in Figure 6.2.

7.4 Search

One of the most common activities we perform when interacting with networks is *search*. Suppose you wish to find some resource that is located on some node of a network. It could be a website with information about a topic of interest, a movie stored on a peer network, or a business contact on a social network — not unlike the target person in Milgram's small-world experiment (Section 2.7). To solve these problems, we need to devise strategies to explore the network efficiently, until the right node is reached. One typically starts from a node of origin and proceeds by visiting neighbors, neighbors of neighbors, and so on. The more effective the strategy, the sooner you can reach the target. This section presents a few prevalent search approaches. In particular, we will emphasize how the peculiar properties of real-world networks can be exploited to expedite the search process.

7.4.1 Local Search

Breadth-first search, presented in Chapter 2, is an attempt to search the whole network by visiting every node, at least in connected components where some seed nodes are known. This type of *exhaustive search* approach solves the problem in some cases, especially when the network is small, or when huge computing and storage resources are available, as demonstrated by Web crawlers that support search engine queries. But often it is more effective, or even necessary, to perform a *local search* of the network (i.e. to perform focused crawls for specific search queries, exploring only a small portion of the network). For instance, you may be interested in a very specific or new piece of Web content that is not listed in a search engine's index. In these cases the search process must employ some heuristics, sorting out the network nodes most likely to contain the desired information.

Another scenario in which local search is necessary is when you wish to download a just-released song from a *peer-to-peer* (or simply *peer*) network, which is a set of personal computers connected directly with each other to share files. Such systems lack a central server that can store the location of every file. This is advantageous because the function of the whole system cannot be compromised by the failure of any single node — say due to a lawsuit or a denial-of-service attack. The disadvantage is that the location of the desired file is unknown. So, whenever users look for a file, queries are sent to the computers of other users that are connected in the peer network. If a computer does not have the requested file, the query is forwarded to one or more neighbors, and so on.

Breadth-first search can also be used for local search, in principle. Starting from the source, we could visit all the nodes of the first layer, and check if any of them is the target node. If not, each of them forwards the query to all their neighbors, and so on, until the right node is reached. Queries already received from other neighbors are ignored. One of the earliest peer networks, called *Gnutella*, used this approach. But breadth-first search is not an efficient strategy. Most of all, it does not take advantage of the structure of the network. In fact, computers on the Gnutella network were flooded with requests and spent all of their bandwidth managing this traffic. That is why Gnutella was eventually replaced by modern peer networks such as *BitTorrent*, that use special network structures designed for efficient search algorithms (Box 7.1).

Box 7.1   **Search in Peer Networks**

Peer networks are used for file sharing and have a structure designed to make the search for shared files efficient. This is achieved by a combination of a *distributed hash table* that maps files to peer computers and an *overlay network* that connects these peer nodes.

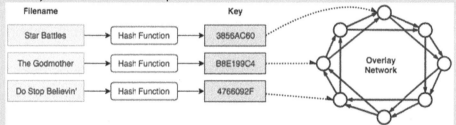

When a file needs to be stored, a unique *key* is generated for the file. This is done by a *hash function*, an algorithm that produces a unique signature from arbitrary data. A key maps to a specific node in the network, so that the file can be routed to that peer. Similarly, when searching for a file, the key is used to forward the query through the network until it reaches the node that has the file with that key. Each node maintains a set of links to its neighbors — a routing table — that is used to forward messages through the overlay network. The distributed hash table of a particular peer network design encodes rules for maintaining the network structure in such a way that search is fast. In particular, for any key, each node either knows the target node that owns that key or has a link to a node that is closer to the target. Thanks to this property, a simple greedy routing algorithm can be employed to forward a message to the neighbor that is closest to the target. Another important property of the peer network is that any computer can join or leave at any time. When a peer leaves or a new peer joins, only the neighbor peers need to be updated; the rest of the network is unaffected.

Fig. 7.16 Model of local search on networks. The source is **s**, the target **t**. The source passes the query to its neighbor with highest degree (**u**), which forwards it to its neighbor of highest degree (**v**). Since the target is a neighbor of **v**, the search ends.

One way to exploit network structure is to rely on the presence of hubs. A local search algorithm based on this idea assumes that each node knows the degree of all of its neighbors as well as the data stored in them, so all information available to nodes is local. When a neighbor of the target node receives the request, it will reply: "I am not the node you are looking for, but my neighbor is!" and send the address of the target node. Each node that is queried, starting from the source, forwards the request to its neighbor with largest degree, unless it or any of its neighbors is the target. The process is repeated until the message is received by a neighbor of the target (Figure 7.16). Since nodes may be visited multiple times during the procedure, those which have passed the request are marked, so that none of them is queried more than once.

In Section 3.3 we have seen that the neighbors of a randomly selected node are more likely to be hubs than the node itself, on average. In particular, by exploring neighbors with large degree, the chance that any of their neighbors is a major hub is higher. Consequently, the algorithm quickly reaches the node with largest degree. After the top hub is checked, it is marked and will be avoided in the future. The next hub is then likely to be the one with the second-largest degree, and so on. Basically, after a rapid transient phase, in which the visited nodes have progressively larger degree, the exploration follows the inverse order of the network's degree sequence, from the node with largest degree downwards. The number of queried nodes, which are the neighbors of the hubs, grows very fast and the target is reached in a small number of steps.

While hub-driven local search brings a gain in the number of steps needed to complete the search, the number of nodes that have to be queried is about the same as when using breadth-first search, on average. This is because the target node can be anywhere, in principle, so many checks are necessary in both cases. The fewer steps of the local search algorithm are compensated by the fact that more neighbors are checked at each step, since the nodes traversed during the procedure have large degree. However, if each node knows the information content of its neighbors, it does not really need to query any of them, which significantly reduces the communication overhead between nodes. This requires hubs to store a huge amount of data, which is unfeasible in very large networks.

7.4.2 Searchability

We have seen a couple of strategies for searching networks. But are all networks "searchable"? Can we search through any graph and expect results in a reasonably short time? The short answer is no, but there are some important exceptions that we discuss next.

To explore the *searchability* properties of a network, recall the small-world experiment by Milgram, presented in Section 2.7. The experiment teaches us two lessons. The first is the familiar observation that most pairs of people in a social network are connected via short chains of acquaintances, as we have seen. The second is that people are surprisingly effective at finding those chains. This is not straightforward: participants knew only their contacts and the name and location of the target person. They had to trust their instinct in the choice of the friend to whom to forward the letter, hoping to get it closer to the target. Most participants tried to send the letter such that it could reach the Boston area as quickly as possible, where the target person lived. This exploits the homophily of the network (discussed in Section 2.1), in particular *geographic homophily*: two people are more likely to know each other if they live nearby. Still, in principle, the letter could have lingered in Boston for a long time once there, being passed among many people before finally reaching the target. Successful participants used some additional intuition about the network structure to find the target in a few steps. They exploited different kinds of homophily based on occupations, say: a lawyer is likely to know another lawyer. This is closely related to topical locality on the Web (Section 4.2.5).

It is possible to analyze the conditions that a network must satisfy in order to be searchable using heuristics based on the types of homophily described above — connecting to a node that is geographically or topically close to the target. Let us first focus on *geographic searchability*. It turns out that there are narrow conditions that make a network geographically searchable. To illustrate this, consider a special structure resembling small-world networks generated by the model discussed in Section 5.2. We start from a square grid, which serves the purpose of embedding the social network in geographic space, like placing people on a map. Each node is connected to its nearest neighbors, forming a grid network. We then add shortcuts between pairs of nodes of the grid (Figure 7.17). At variance with the small-world model [Figure 5.4(b)], the shortcuts do not connect pairs of nodes with equal probability; rather, the link probability decreases with the geographic distance between the nodes in the grid. This is designed to account for geographic homophily, the empirical observation that most relationships in real social networks occur among people in geographic proximity to each other.

Let us assume that each individual knows exactly the geographic position of their neighbors, as well as the position of the target. Therefore each individual can determine precisely which neighbor is geographically closest to the target. For the sake of simplicity, let us further assume that the source and target nodes are chosen at random, and that people follow the *greedy search algorithm* inspired by Milgram's experiment: each node forwards the message across a link that brings it as close as possible to the target. We can define the *delivery time* as the number of times the message is passed between nodes until it reaches the target. As it turns out, the delivery time is very short only if the shortcut probability falls off in just the right way as a function of the geographic distance between nodes. In the case of a two-dimensional grid as shown in Figure 7.17, the probability of a shortcut must decay as the inverse of the square of the distance. For example, a link between two nodes lying two steps apart from each other should be four times more likely than a link connecting two nodes that are twice as far (four steps).

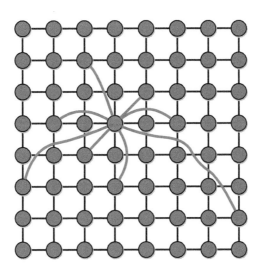

Fig. 7.17 Geographic social network. The square grid represents the geographic area where the people (nodes) live. Each node is linked to its four nearest neighbors. Shortcuts between the nodes are added favoring pairs of individuals living close to each other. The figure only shows the shortcuts of the red node.

If the shortcut probability falls more rapidly with the distance between the nodes, there are not enough long-range links, so one is doomed to traverse many local links before reaching the target. If the shortcut probability falls more slowly, there are too many long-range links. In this scenario there are many short paths, but they are hard to find, like searching for a needle in a haystack. In both cases, the search process is not very efficient and the greedy search algorithm needs a long time to find the target.

Although the condition for geographic searchability of a network is quite narrow in this scenario, it is not entirely unrealistic. In the Web, if we replace the notion of geographic homophily with that of topical locality, we can measure empirically the probability that two pages are linked as a function of their topical distance. Imagine that the grid of Figure 7.17 represents a topical landscape, and that nearby points represent related Web pages. In practice, we can measure the similarity between two pages by looking at their content (recall Box 4.1). Small similarity values can be mapped to large distances and vice versa. It turns out that nearby (similar) pages are highly likely to have common neighbors or be linked, while for distant (dissimilar) pages the decay in link probability is compatible with the geographic searchability condition. The Web is therefore a special case of a searchable network, which is reassuring as it means we can find interesting information by clicking on links. If this were not the case, surfing the Web would be hopeless.

The network model used to explore geographic searchability is unrealistic in many ways. People are not located and connected as the nodes of a grid. More importantly, geography is just one of the many possible attributes of the nodes in a network being searched. Two people in a social network may have the same job, practice the same hobby, attend the same school, and so on. Let us generalize the notion of searchability to *topical searchability*, where any attribute of the nodes can be reflected in network homophily and thus facilitate

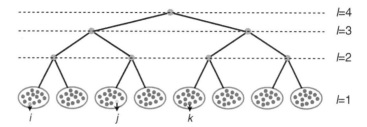

Fig. 7.18 Topical distance tree. The distance between nodes i and j is three, because the nearest common ancestor of the groups to which i and j belong is at level three (green dot on the left of the $l = 3$ dashed line). Likewise, the topical distance between i and k or j and k is $l = 4$ because the nearest common ancestor is the root.

the search process. For instance, as mentioned earlier, the occupation of the target node was a useful piece of information in Milgram's experiment.

We can group the nodes of a network in a hierarchical manner based on their topical attributes: the top of the hierarchy represents the most general category, which is split into smaller, increasingly specific topical categories as we go down, until we reach the smallest groups that we can identify. The resulting hierarchical diagram is a *topical distance tree*, illustrated in Figure 7.18. A topical distance tree could be used to organize Wikipedia articles about science. At the top (below the root) would be the formal sciences, physical sciences, life sciences, social sciences, and applied sciences. In the level below we would find disciplines like math, logic, biology, chemistry, physics, psychology, economics, sociology, engineering, computer science, and so on. More specific fields, such as molecular biology, statistical physics, machine learning, and network science would be placed at lower levels. Similarly, one can use a topical tree to classify people in a social network. At the top would be the whole-world population and the lower groups could represent a geographic subdivision of the population into continents, countries, cities, and neighborhoods. Different social attributes (e.g. occupations, hobbies, schools, religions) lead to different divisions and trees.

A topical distance tree is a mental construct that allows us to estimate the *topical distance* between nodes (Figure 7.18). If two individuals belong to the same smallest identifiable group, their topical distance is one. This would be the case, say, for two professors working in the same department at Indiana University in Bloomington. Otherwise, their groups will eventually merge as we climb up the hierarchical tree. This happens when we bump into their *nearest-ancestor* category in the tree, which represents the most specific attribute shared by the nodes. In this case, the topical distance is given by the number of levels in the tree, from the bottom up to the nearest common ancestor. For instance, in the schematic diagram of Figure 7.18, individuals i and j could be two professors working in distinct departments of different universities in Indiana, so their topical distance is three, because working on separate topics and in different locations adds two degrees of separation.

Let us stick with the social network scenario and assume that people can estimate their topical distance from anybody. This is a less stringent hypothesis than in the geographic model, where individuals know each other's exact position. Let us further assume that the topical distance tree captures the social network's homophily, so that the link probability between two nodes decreases as their topical distance increases, according to a decay function. By using the greedy search algorithm (i.e. by letting each person forward the message to the neighbor with the shortest topical distance from the target) it can be shown that there is a special topical decay function that allows for efficient search. In this condition the search takes a small number of steps.

The condition for topical searchability of the network, expressed by the relationship between topical distance and link probability, is quite strict. However, it is sociologically plausible, helping us understand the successful chains in Milgram's experiment. Furthermore, one can measure how the probability that two Web pages are linked decays with their topical distance by analyzing pages classified in a topical Web directory. It turns out that the Web graph meets the topical searchability condition as well, confirming that it is searchable by surfing.

7.5 Summary

Networks are vehicles for the diffusion of ideas, opinions, and influence. By the same token, they facilitate harmful spreading processes, like the diffusion of infections, misinformation, and rumors. Uncovering how these phenomena unfold can help us improve the effectiveness of the former and defend ourselves from the latter. Searching networks is critical for retrieving information, but difficult when the network structure and the content stored by the nodes are unknown. In this chapter we have reviewed simple models describing these processes and learned the following key lessons:

1. In threshold models of influence diffusion, a node/individual is subject to the combined effect of all its neighbor influencers: when this effect exceeds a threshold, the node is affected. In independent cascade models, a node/individual is "convinced" by each neighbor influencer with a certain probability. The most effective influencers have large degree and a central position in the network.
2. In the susceptible–infected–susceptible (SIS) model of epidemic spreading, when infected individuals recover they become susceptible again, so they can contract the disease multiple times. In the susceptible–recovered–susceptible (SIR) model, when infected individuals recover they cannot be infected anymore, so they play no further role in the dynamics.
3. If the contact networks have hubs, a disease spreading according to both SIR and SIS dynamics can affect an important fraction of the population, even if the probability of infection is low, because the hubs can easily be infected and turn into dangerous spreaders.
4. The rumor-spreading model is similar to SIR, but the "recovery" process, corresponding to the decision to not spread the rumor further, is a consequence of encounters between individuals who know the rumor, instead of happening spontaneously for each individual. The rumor can reach a significant portion of any network, even for low transmission probability.

5. In the majority opinion model, a node takes the opinion of the majority of its neighbors. Different opinions coexist in the final state. Consensus is only reached on one- and two-dimensional grids; in these cases, the consensus opinion is the majority opinion in the initial configuration.

6. In the voter model, a node takes the opinion of a randomly selected neighbor. The dynamics lead to consensus on all networks. Consensus on an opinion is reached with a probability that matches the fraction of nodes holding that opinion in the initial configuration.

7. In bounded-confidence models of continuous opinion dynamics, two opinions can affect each other only if their difference is smaller than the confidence bound parameter. The final number of opinion clusters depends on the value of the confidence bound and the network structure. With a sufficiently large confidence bound, the dynamics lead from random initial opinions to consensus on any network.

8. Coevolution models combine the processes of selection and social influence. We presented a model in which a node can either take the opinion of a neighbor or select a new neighbor with the same opinion. In the final state, the system is segregated into homogeneous opinion communities, disconnected from each other.

9. For an exhaustive search of a network, like those performed by Web crawlers, the standard approach is breadth-first search, the same algorithm used to compute distances and find shortest paths between nodes. This can be unfeasible for large networks, so that local heuristic search becomes necessary. One local search heuristic is to forward the query to the neighbor nodes with largest degree, so that we can quickly reach the biggest hubs and exploit their large numbers of neighbors to find the target in a small number of steps.

10. Some networks are searchable, in that one can find short paths that connect a source to a target. This may be due to a peculiar geographic distribution of links between nodes, or to a hierarchical organization of nodes according to their content or attributes. By estimating the distance between two nodes in the hierarchy, one can identify the neighbor closest to the target.

7.6 Further Reading

Most general books on network science, like the ones recommended in Section 1.12, have ample sections on dynamic processes. The book by Barrat *et al.* (2008) is dedicated to the topic, and covers in detail most of the models presented in this chapter.

The science behind the spread of misinformation is an emerging area of research (Lazer *et al.*, 2018). The study of information diffusion networks (Shao *et al.*, 2018a) is critical to help us understand how social media can be manipulated, for example via social bots (Shao *et al.*, 2018b).

Threshold models were introduced in a classic paper by Granovetter (1978), while the independent cascade model is more recent (Goldenberg *et al.*, 2001). Watts (2002) proposed imposing a threshold on the fraction of neighbors, instead of their number. Kempe *et al.* (2003) addressed the problem of identifying the set of influencers who can generate the largest cascades. Kitsak *et al.* (2010) showed that the hubs are not necessarily the most effective influencers. Centola and Macy (2007) explored complex contagion in the spread of collective behaviors. Weng *et al.* (2013b) showed that communities affect the viral spread of memes in social media, and that cascade sizes can be predicted based on how many communities are involved in the early stages of diffusion.

The book by Anderson and May (1992) is a good reference for classic epidemic modeling. Pastor-Satorras *et al.* (2015) published a comprehensive review of epidemic processes on networks. Stehlé *et al.* (2011) reconstructed the network of face-to-face interactions among children and teachers in a school, by means of radio-frequency identification devices. The lack of an epidemic threshold on networks with hubs was first exposed by Pastor-Satorras and Vespignani (2001). Cohen *et al.* (2003) suggested that immunizing the acquaintances of randomly selected individuals is an effective strategy if the networks of contacts have heavy-tailed degree distributions. Christakis and Fowler (2010) showed that monitoring the friends of randomly selected individuals allows early detection of epidemic outbreaks. The rumor-spreading model was first presented by Daley and Kendall (1964).

Castellano *et al.* (2009) review opinion and other social dynamics models from the point of view of statistical physics. The majority model was originally introduced in the context of spin models in statistical physics (Glauber, 1963). Another model based on the concept of majority, not discussed in this chapter, is called the *majority rule model* (Galam, 2002; Krapivsky and Redner, 2003). The voter model was proposed to describe the territorial competition among species (Clifford and Sudbury, 1973). Mobilia *et al.* (2007) studied the role of zealots in the voter model. Vázquez *et al.* (2003b) developed the *constrained voter model*, in which only similar opinions can interact. The principle of bounded confidence dates back to Festinger's (1954) theory of social comparison. The original bounded-confidence opinion model was introduced by Deffuant *et al.* (2000). The first models of coevolution of network dynamics and structure were proposed by Holme and Newman (2006) and Gil and Zanette (2006).

Adamic *et al.* (2001) proposed the local search strategy that relies on the presence of hubs in the network. The geographic network and the relative analysis of network searchability were presented by Kleinberg (2000). Analyses of searchability based on topical hierarchies and distances were put forward independently by Kleinberg (2002) and Watts *et al.* (2002). Menczer (2002) showed that the Web graph satisfies versions of geographic and topical searchability.

Exercises

7.1   Go through the Chapter 7 Tutorial on the book's GitHub repository.[1] It provides a class that makes it easy to code and run simulations of network dynamics models.

---

[1] github.com/CambridgeUniversityPress/FirstCourseNetworkScience

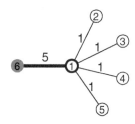

A weighted influence network. Node **1** has only one active neighbor (**6**).

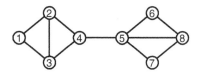

An influence network. Each node has a threshold of 1/2.

7.2   Consider the example in Figure 7.19. According to the linear threshold model, will node **1** be activated if its threshold is 4? What if it is 5? Do the answers to these questions change if we vary the weights of the links joining node **1** to its inactive neighbors?

7.3   Someone gives you a network with some of its nodes activated. She claims that you will never succeed in activating all the nodes, no matter which model of influence spreading you use. How can she be so sure?

7.4   Apply the fractional threshold model to the network in Figure 7.20. The threshold is 1/2 for all nodes. Which node should we activate to obtain the largest cascade? Is the solution unique? What is the minimum number of initial influencers that are needed to activate the whole network?

7.5   You are considering the independent cascade model on a network. Two active nodes **s** and **t** have degree 4 and 10, respectively. They can convince their neighbors with probability $1/2$ (**s**) and $1/5$ (**t**). Which node will influence more neighbors on average, **s** or **t**?

7.6   Consider the network in Figure 7.21: the influence probabilities are symmetric, for example the probability that node **1** convinces node **2** is equal to the probability that node **2** convinces node **1**. Use the independent cascade model to predict how many nodes will be active in the end, on average, by activating node **2** initially.

7.7   Contagion models such as SIS and SIR come from epidemiology, but as it turns out, they can model other spreading processes on networks quite well. Which of the following processes could best be described by an SIS model on a network?
a. The spread of toxic gas through the air over a geographic region
b. The spread of an oil slick over the surface of a body of water

Fig. 7.21 A network with symmetric influence probabilities shown next to the links. Node **2** is active.

      c. The impact of a power station failure in the US power grid

      d. The adoption of a specific smartphone among members of a community

7.8   The game Pandemic II (pandemic2.org) is based on an elaborate SIR model. Play the game and write a brief report on how the various aspects of the game correspond to SIR model mechanisms. Discuss key simplifying assumptions made in the game. Describe how various game choices affect the model parameters.

7.9   Consider SIS model dynamics on a population. Suppose that a fraction f of the population never gets sick and that such immune individuals are randomly distributed in a homogeneous contact network (all nodes have similar degree). Is the risk of epidemic spreading bigger or smaller than in the pure SIS model, for which $f = 0$? Would the answer change if we considered SIR instead? [*Hint*: Use the condition of Eq. (7.5).]

7.10 There is an epidemic outbreak and after a quick verification it turns out that the basic reproduction number is $R_0 = 2.5$, so we are heading towards an epidemic spread (assume that the contact network is homogeneous). The authorities urge the population to limit their contact with other people, so that, on average, each individual gets in touch with about half the usual number of people. Suppose that doctors are capable of developing medicines that can significantly increase the recovery rate μ. How much does μ have to increase so that the epidemic can be stopped?

7.11 Simulate the SIR dynamics on a random network with $N = 1000$ nodes and link probability $p = 0.01$. Initially 10 nodes are infected, chosen at random. The probability of recovery is $\mu = 0.5$. Run the dynamics for these values of the infection probability: $\beta = 0.02, 0.05, 0.1, 0.2$. In each run, save the number of simultaneously infected people after each iteration and calculate the maximum value. Interpret the results. How many iterations are needed to reach the maximum? Do you observe a major outbreak? Why or why not? (*Hint*: Feel free to modify the code in this chapter's tutorial to run the simulations.)

7.12 In a community there are three kinds of people: frustrated (S), aggressive (I), and peaceful (R). When a frustrated individual meets an aggressive one, they become aggressive with probability β. When an aggressive individual meets a peaceful one, they become peaceful with probability α. When two aggressive people meet each other, they start to argue. But with probability α, they realize after a while that fighting is futile, and so they both turn peaceful. Can a major spread in aggressive behavior be prevented by a small value of β?

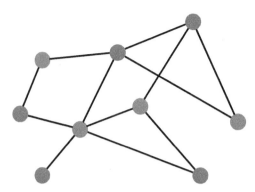

Fig. 7.22   A network with nodes colored red or blue according to their opinions.

7.13 In the network illustrated in Figure 7.22, each node has one of two possible opinions. An *active link* connects nodes in different opinion states. These links are called active because in theory either endpoint has a chance of convincing the other to adopt its opinion, depending on the specific rules of the model. How many active links are there?

7.14 When we simulate a dynamic process on a network, there are several ways to asynchronously pick the next node(s) to update. Typically, nodes are selected in random sequence. Another strategy would be to select one endpoint of a randomly selected link. Do you think that this would affect the dynamics in any way? Why or why not?

7.15 Simulate the majority opinion dynamics on a square grid with $N = 20 \times 20 = 400$ nodes. Initially assign each of two opinions to half of the nodes chosen at random. Execute 100 runs with different initial random assignments, until the system gets to a stationary state. How many runs lead to consensus? Create a histogram of the proportion of opinion one in the non-consensus stationary states. Create a histogram of the fraction of active links in those configurations. (Active links are defined in Exercise 7.13. The fraction of active links is the ratio between the number of active links and the total number of links in the network.) [Hint: If you use the code in this chapter's tutorial to run the simulations, to guarantee convergence to a stationary state you have to write the `state_transition()` function in such a way that nodes are updated in asynchronous fashion and random order. You also have to specify a stop condition function to end the simulation when the stationary state is reached.]

7.16 Compute the exit probability of the majority opinion model on a square grid with $N = 20 \times 20 = 400$ nodes. Let the initial fraction of nodes with opinion one be $p = 0.1, 0.2, 0.3, 0.4, 0.5, 0.6, 0.7, 0.8, 0.9$. Execute 20 runs with different initial random assignments for each value of p, until the system gets to a stationary state. Consider only the runs leading to consensus, and for each p compute the fraction of those runs for which consensus is on opinion one, which is the exit probability for that value of p. Plot the result as a function of p. (*Hint*: Feel free to modify the code in this chapter's tutorial to run the simulations, as explained in the previous exercise.)

7.17 Compute and plot the exit probability of the voter model on the square grid. Use the same parameters as in Exercise 7.16. (*Hint*: Feel free to modify the code in this chapter's tutorial to run the simulations, as explained in the previous exercises.)

7.18 Consider the bounded-confidence model of opinion dynamics on a complete network. Since all nodes are connected to each other, any two nodes can affect each other's opinion if they are sufficiently close. A mathematical argument shows that if the initial opinions are randomly distributed in the interval $[0, 1]$, the number of final opinion clusters in this case is approximately equal to $\frac{1}{2\epsilon}$, where ϵ is the confidence bound. If you are mathematically inclined, can you give an intuition into this argument?

7.19 Simulate the bounded-confidence model of opinion dynamics on a complete network with $N = 1000$ nodes. The initial opinions are random numbers between zero and one. Consider three different values of the confidence bound: $\epsilon = 0.125, 0.25, 0.5$. For each ϵ, use different values for the convergence parameter, say $\mu = 0.1, 0.3, 0.5$. Run every simulation until each opinion varies by less than 1% between consecutive iterations, and plot a histogram of the final opinions. Does the number of final opinion clusters depend on ϵ? Why or why not? Does it depend on μ? Why or why not? (*Hint*: Feel free to modify the code in this chapter's tutorial to run the simulations.)

7.20 Simulate the bounded-confidence model of opinion dynamics on a random network with $N = 1000$ nodes and link probability $p = 0.01$. The initial opinion configuration is generated by assigning to each node a random number between zero and one. Set the parameter $\mu = 1/2$ and explore different values of the confidence bound ϵ. Run every simulation until each opinion varies by less than 1% between consecutive iterations. What is the threshold ϵ_c such that, for $\epsilon > \epsilon_c$, we have a single opinion cluster (consensus) in the final configuration? Now simulate the model on a small-world network with $N = 1000$ nodes, $k = 4$, and rewiring probability $p = 0.01$. What is ϵ_c in this case? (*Hint*: Feel free to modify the code in this chapter's tutorial to run the simulations.)

7.21 Consider the coevolution model with just two opinions, initially distributed randomly among the nodes. How many opinion communities do you expect there will be when selection dominates (p close to 1)? What is their size, approximately? (*Hint*: You may assume that the network is not too sparse.)

7.22 In the coevolution model, the influence component follows the rule of the voter model, in that the node takes the opinion of a random neighbor. Let us see what happens if we switch to majority dynamics. The new model works like this: given a node, with probability p it rewires one of its links to a non-neighbor node with the same opinion, like before; with probability $1 - p$ it takes the majority opinion of its neighborhood. Describe the final configurations you expect to observe when the system reaches the stable state in the extreme cases of p close to zero and one.

7.23 Build small-world networks with $N = 1000$ nodes, $k = 4$, and these values for the rewiring probability: $p = 0.001, 0.01, 0.1, 1$. Choose a source node **s** and a target

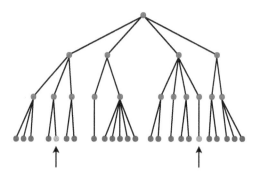

Fig. 7.23 A topical distance tree.

node **t** at random. Apply the greedy search algorithm, where the message is passed to the neighbor that is closest to the target along the ring, and compute the number of steps needed to deliver the message from **s** to **t** for each value of p. Interpret the results. (*Hint*: For each p, average your measurement across multiple runs with different random pairs of nodes.)

7.24 The topical distance tree in Figure 7.18 is very stylized and unrealistic. Real topical distance trees are generally asymmetric, like the one in Figure 7.23. What is the topical distance between the two marked individuals?

Appendix A: **Python Tutorial**

This tutorial demonstrates the features of the Python programming language and of Jupyter Notebooks, which are used in the examples and tutorials of the textbook. Special attention is given to the most important data types used in data analysis workflows, as well as common idioms and patterns employed in these use cases. This appendix may be especially useful for readers more experienced in programming languages other than Python.

This tutorial, along with tutorials for each chapter, are available as Jupyter Notebooks on the book's GitHub repository.[1]

A.1 Jupyter Notebook

The Jupyter Notebook is an open-source Web application. There are free Jupyter environments that require no setup and run entirely in the cloud, such as Google's Colaboratory.[2] You can also install Jupyter Notebook with Anaconda,[3] a free distribution that includes Python, Jupyter, NetworkX, and other commonly used packages for scientific computing and data science.

Even those well versed in Python may not have used a Jupyter Notebook before. The main idea is that we can mix text and code, similar to the presentation in this book, and that code is executed in "cells." By clicking on a cell and pressing Shift + Enter, you execute the cell and move to the next cell. Ctrl + Enter executes the cell but does not move to the next cell. You can run many cells at once by using the different options in Jupyter's "Cell" menu.

The code snippets presented in this appendix are organized like Jupyter Notebook cells. Each outlined section is a cell; the output, if any, of executing the cell's code is printed below the dark red separator, as seen in this simple example:

```
print('Hello from Jupyter')
```
```
Hello from Jupyter
```

---

[1] github.com/CambridgeUniversityPress/FirstCourseNetworkScience
[2] colab.research.google.com
[3] www.anaconda.com

In Jupyter Notebooks, we have two different ways of inspecting variables. Python's `print()` function is useful as always:

```
my_str = 'Hello'
my_int = 16

print(my_str)
print(my_int)
```
```
Hello
16
```

We can also just execute a cell with the name of a variable:

```
my_str
```
```
'Hello'
```

The big difference here between the two approaches is that `print()` statements can output multiple items per cell, while the latter approach will only display the last variable named. Observe:

```
my_str
my_int
```
```
16
```

As opposed to the first example using `print()`, this only outputs the last value.

A.2 Conditionals

"Conditionals" is a fancy word for if statements. If you've ever done any programming, you are doubtless aware of the if–then–else construction. In Python it's done as follows:

```
number_of_apples = 5

if number_of_apples < 1:
    print('You have no apples')
elif number_of_apples == 1:
    print('You have one apple')
elif number_of_apples < 4:
```

```
    print('You have a few apples')
else:
    print('You have many apples!')
```
```
You have many apples!
```

You can change `number_of_apples` and re-run the previous cell in order to get the different possible outputs.

A.3 Lists

One of Python's most versatile and ubiquitous data types is "List." This is an *ordered, mutable collection* of *non-unique* items.

By *ordered*, we mean that the items are addressed by their *index* in the collection:

```
student_names = ['Alice', 'Bob', 'Carol', 'Dave']
student_names[1]
```
```
'Bob'
```

Indices in Python start at zero, so the head of the list has index 0:

```
student_names[0]
```
```
'Alice'
```

We can get the last item in a list by using negative indexing:

```
student_names[-1]
```
```
'Dave'
```

Lists can also be *sliced* to get a subset of the list items:

```
student_names[0:2]
```
```
['Alice', 'Bob']
```

```
student_names[1:3]
```
```
['Bob', 'Carol']
```

When slicing from the beginning of the list, or to the end of the list, we can leave out the index:

```
student_names[:2]

['Alice', 'Bob']
```

```
student_names[2:]

['Carol', 'Dave']
```

By *mutable*, we mean that the list can be changed by adding or removing items. We most often add items to the end of the list with `.append()`:

```
student_names.append('Esther')
student_names

['Alice', 'Bob', 'Carol', 'Dave', 'Esther']
```

But we can also add items at any arbitrary index with `.insert()`:

```
student_names.insert(2, 'Xavier')
student_names

['Alice', 'Bob', 'Xavier', 'Carol', 'Dave', 'Esther']
```

We can delete items with the `del` keyword:

```
del student_names[2]
student_names

['Alice', 'Bob', 'Carol', 'Dave', 'Esther']
```

List items do *not* have to be *unique*. Nothing stops us from repeatedly adding the same name to this list:

```
student_names.append('Esther')
student_names

['Alice', 'Bob', 'Carol', 'Dave', 'Esther', 'Esther']
```

If you want a collection where uniqueness is enforced, you should look towards dictionaries or sets.

A *collection* refers to a data type consisting of more than one values. Lists are one type of collection, but there are others such as tuples, sets, and dictionaries.

When naming your variables that contain lists, it is a helpful convention to use plural nouns, such as `student_names` in the previous example. In contrast, single values should be named with singular nouns, like `my_str` in the first section. This helps you and others reading your code keep straight which variables are collections and which are single items, and also helps when writing loops, as shown in the next section.

A.4 Loops

If you're coming from another programming language, you're probably aware of more than one type of loop. In Python, we focus on one type of loop in particular: the for loop. The for loop iterates through a collection of items, executing its code for each item:

```
student_names = ['Alice', 'Bob', 'Carol', 'Dave']

for student_name in student_names:
    print('Hello ' + student_name + '!')
```
```
Hello Alice!
Hello Bob!
Hello Carol!
Hello Dave!
```

Note the naming convention being used in the for–in construction:

```
for student_name in student_names:
```

By using a plural noun for the collection `student_names`, we automatically have a good name for each individual item in the collection: `student_name`. The tutorials in this book use this naming convention when possible, as it makes clear to the reader which variable is the "loop variable" that changes value between iterations of the loop body.

A common type of task when working with data is the *filtering task*. In abstract, this task involves looping over a collection, checking each item for some criterion, then adding items that meet the criterion to another collection.

In the following example, we'll create a list of just the "long" names from the `student_names` list. Long names are those that contain more than four characters. You will often see and write code that looks like the following in this book's tutorials:

```
# Initialize an empty list and add to it the
# student names containing more than four characters
```

```
long_names = []
for student_name in student_names:
    # This is our criterion
    if len(student_name) > 4:
        long_names.append(student_name)

long_names
```
```
['Alice', 'Carol']
```

Loops can be "nested" inside one another. This often occurs when we want to match up items from one collection to items from the same or another collection. Let's create a list of all possible pairs of students:

```
student_names = ['Alice', 'Bob', 'Carol', 'Dave']

student_pairs = []
for student_name_0 in student_names:
    for student_name_1 in student_names:
        student_pairs.append(
            (student_name_0, student_name_1)
        )

student_pairs
```
```
[('Alice', 'Alice'),
 ('Alice', 'Bob'),
 ('Alice', 'Carol'),
 ('Alice', 'Dave'),
 ('Bob', 'Alice'),
 ('Bob', 'Bob'),
 ('Bob', 'Carol'),
 ('Bob', 'Dave'),
 ('Carol', 'Alice'),
 ('Carol', 'Bob'),
 ('Carol', 'Carol'),
 ('Carol', 'Dave'),
 ('Dave', 'Alice'),
 ('Dave', 'Bob'),
 ('Dave', 'Carol'),
 ('Dave', 'Dave')]
```

Note here that instead of just adding names to the student_pairs list, we are adding *tuples* (student_name_0, student_name_1). This means each item in the list is a 2-tuple:

```
student_pairs[0]
```
```
('Alice', 'Alice')
```

We'll talk more about tuples in the next section. The second thing to notice is that we're including pairs with two of the same student. Suppose we wish to exclude those. We can accomplish this by adding an if statement in the second for loop to *filter* out those repeats:

```
student_names = ['Alice', 'Bob', 'Carol', 'Dave']

student_pairs = []
for student_name_0 in student_names:
    for student_name_1 in student_names:
        # This is the criterion we added
        if student_name_0 != student_name_1:
            student_pairs.append(
                (student_name_0, student_name_1)
            )

student_pairs
```
```
[('Alice', 'Bob'),
 ('Alice', 'Carol'),
 ('Alice', 'Dave'),
 ('Bob', 'Alice'),
 ('Bob', 'Carol'),
 ('Bob', 'Dave'),
 ('Carol', 'Alice'),
 ('Carol', 'Bob'),
 ('Carol', 'Dave'),
 ('Dave', 'Alice'),
 ('Dave', 'Bob'),
 ('Dave', 'Carol')]
```

And now the list has no repeats.

A.5 Tuples

Even experienced Python users are often confused about the difference between tuples and lists, so definitely read this short section even if you have some experience.

Tuples are superficially similar to lists as they are ordered collections of non-unique items:

```
student_grade = ('Alice', 'Spanish', 'A-')
student_grade
```

```
('Alice', 'Spanish', 'A-')
```

```
student_grade[0]
```

```
'Alice'
```

The big difference from lists is that tuples are *immutable*. Each of the following cells should raise an exception:

```
student_grade.append('IU Bloomington')
```

```
Traceback (most recent call last):
  <ipython-input-24-782d93a0b0cf> in <module>()
  ----> 1 student\_grade.append('IU Bloomington')

  AttributeError: 'tuple' object has no attribute 'append'
```

```
del student_grade[2]
```

```
Traceback (most recent call last):
  <ipython-input-25-f8ded3b186ff> in <module>()
  ----> 1 del student\_grade[2]

  TypeError: 'tuple' object doesn't support item deletion
```

```
student_grade[2] = 'C'
```

```
Traceback (most recent call last):
  <ipython-input-26-c9fd9c464431> in <module>()
  ----> 1 student\_grade[2] = 'C'

  TypeError: 'tuple' object does not support item assignment
```

This immutability makes tuples useful when *indices matter*. In this example, the index matters semantically: index 0 is the student's name, index 1 is the course name, and index 2 is their grade in the course. The inability to insert or append items to the tuple means we are certain that, say, the course name won't move to a different position.

Tuples' immutability makes them useful for *unpacking*. At its simplest, tuple unpacking allows the following:

```
student_grade = ('Alice', 'Spanish', 'A-')
student_name, subject, grade = student_grade

print(student_name)
print(subject)
print(grade)
```
```
Alice
Spanish
A-
```

Tuple unpacking is most useful when used with loops. Consider the following piece of code, which congratulates students on getting good grades:

```
student_grades = [
    ('Alice', 'Spanish', 'A'),
    ('Bob', 'French', 'C'),
    ('Carol', 'Italian', 'B+'),
    ('Dave', 'Italian', 'A-'),
]

for student_name, subject, grade in student_grades:
    if grade.startswith('A'):
        print('Congratulations', student_name,
              'on getting an', grade,
              'in', subject)
```
```
Congratulations Alice on getting an A in Spanish
Congratulations Dave on getting an A- in Italian
```

Compare this to the same code using indices:

```
for student_grade in student_grades:
    if student_grade[2].startswith('A'):
        print('Congratulations', student_grade[0],
```

```
        'on getting an', student_grade[2],
        'in', student_grade[1])
```
```
Congratulations Alice on getting an A in Spanish
Congratulations Dave on getting an A- in Italian
```

Tuple unpacking allows us to easily refer to this structured data by semantic names instead of having to keep the indices straight. The second example, while functionally identical, is more difficult to write and harder still to read.

A.6 Dictionaries

The next type of collection is different from the previous two, but is among the most powerful tools in Python: the dictionary. The dictionary is an *unordered*, *mutable* collection of *unique* items. In other programming languages, dictionaries are called maps, mappings, hashmaps, or associative arrays.

By *unordered*, we mean that dictionary items aren't referred to by their position, or index, in the collection. Instead, dictionary items have *keys*, each of which is associated with a value. Here's a very basic example:

```
foreign_languages = {
    'Alice': 'Spanish',
    'Bob': 'French',
    'Carol': 'Italian',
    'Dave': 'Italian',
}
```

Here the student names are the keys and the languages are the values. So to see Carol's language, we use the key — her name — instead of an index:

```
foreign_languages['Carol']
```
```
'Italian'
```

Trying to get the value for a key that does not exist in the dictionary results in a KeyError:

```
foreign_languages['Zeke']
```
```
Traceback (most recent call last):
  <ipython-input-32-1ff8fc89736a> in <module>()
```

```
----> 1 foreign\_languages['Zeke']

KeyError: 'Zeke'
```

We can check if a particular key is in a dictionary with the in keyword:

```
'Zeke' in foreign_languages

False
```

```
'Alice' in foreign_languages

True
```

Note that keys are case sensitive:

```
'alice' in foreign_languages

False
```

We can add, delete, and change entries in a dictionary:

```
# Add an entry that doesn't exist
foreign_languages['Esther'] = 'French'
foreign_languages

{'Alice': 'Spanish',
 'Bob': 'French',
 'Carol': 'Italian',
 'Dave': 'Italian',
 'Esther': 'French'}
```

```
# Delete an entry that exists
del foreign_languages['Bob']
foreign_languages

{'Alice': 'Spanish',
 'Carol': 'Italian',
 'Dave': 'Italian',
 'Esther': 'French'}
```

```
# Change an entry that does exist
foreign_languages['Esther'] = 'Italian'
foreign_languages
```
```
{'Alice': 'Spanish',
 'Carol': 'Italian',
 'Dave': 'Italian',
 'Esther': 'Italian'}
```

Note that the syntax for adding an entry that does not exist and changing an existing entry are the same. Assigning a value to a key in a dictionary adds the key if it doesn't exist, or else updates the value for the key if it does exist. As a consequence, keys are necessarily *unique* — there can't be more than one key with the same name in a dictionary.

It is possible to loop over entries in a dictionary. Here is one way to accomplish this task:

```
for student, language in foreign_languages.items():
    print(student, 'is taking', language)
```
```
Alice is taking Spanish
Carol is taking Italian
Dave is taking Italian
Esther is taking Italian
```

In `foreign_languages` we have paired data — every name is associated with a subject. Dictionaries are also often used to contain several different data about a single entity. To illustrate this subtle difference, let's take a look at one item from `student_grades`:

```
student_grade = ('Alice', 'Spanish', 'A')
```

Here we know that the items in each of these tuples are a name, subject, and grade:

```
student_name, subject, grade = student_grades[0]
print(student_name,
      'got a grade of', grade,
      'in', subject)
```
```
Alice got a grade of A in Spanish
```

We could instead represent this data as a dictionary. A dictionary of information describing a single item is often referred to as a *record*:

```
record = {
    'name': 'Alice',
    'subject': 'Spanish',
    'grade': 'A',
}
print(record['name'],
        'got a grade of', record['grade'],
        'in', record['subject'])
```
```
Alice got a grade of A in Spanish
```

While the code is slightly longer, there is absolutely no ambiguity here about matching up indices and what each value represents. This is also useful in situations where some of the fields might be optional.

A.7 Combining Data Types

In most of these simple examples we've worked with collections of simple values like strings and numbers, however data analysis often involves working with complex data, where each item of interest has several data types associated with it. This complex data is often represented as collections of collections, for example lists of dictionaries.

Choosing the appropriate data types for a given problem will make it easier for you to write bug-free code and will make your code easier for others to read, but identifying the best data types is a skill gained through experience. A few commonly used combination data types are illustrated below, but this is hardly an exhaustive list.

A.7.1 List of Tuples

We've actually seen this one before. Consider the student_grades data from the earlier example on tuple unpacking:

```
student_grades = [
    ('Alice', 'Spanish', 'A'),
    ('Bob', 'French', 'C'),
    ('Carol', 'Italian', 'B+'),
    ('Dave', 'Italian', 'A-'),
]
```

This is a list of tuples:

```
student_grades[1]
```
```
('Bob', 'French', 'C')
```

and we can work with the individual tuples:

```
student_grades[1][2]
```
```
'C'
```

A.7.2 List of Dictionaries

In the section on dictionaries, we explored how a dictionary is often used to contain a record about a single entity. Let's convert the list of tuples `student_grades` into a list of records `student_grade_records`:

```
student_grade_records = []
for student_name, subject, grade in student_grades:
    record = {
        'name': student_name,
        'subject': subject,
        'grade': grade,
    }
    student_grade_records.append(record)

student_grade_records
```
```
[{'name': 'Alice', 'subject': 'Spanish', 'grade': 'A'},
 {'name': 'Bob', 'subject': 'French', 'grade': 'C'},
 {'name': 'Carol', 'subject': 'Italian', 'grade': 'B+'},
 {'name': 'Dave', 'subject': 'Italian', 'grade': 'A-'}]
```

Now each item in the list is a dictionary:

```
student_grade_records[1]
```
```
{'name': 'Bob', 'subject': 'French', 'grade': 'C'}
```

and we can work with the individual records:

```
student_grade_records[1]['grade']
```

```
'C'
```

This list of dictionaries is often used to represent data from a database or an API. Let's use this data to write our code congratulating students for good grades, as we did in the section on tuple unpacking:

```
for record in student_grade_records:
    if record['grade'].startswith('A'):
        print('Congratulations', record['name'],
              'on getting an', record['grade'],
              'in', record['subject'])
```

```
Congratulations Alice on getting an A in Spanish
Congratulations Dave on getting an A- in Italian
```

A.7.3 Dictionary of Dictionaries

The list of dictionaries is very useful when dealing with non-unique data; in the previous example each student might have several grades from different classes. But sometimes we want to refer to the data by a particular name or key. In this case, we can use a dictionary whose values are records (i.e. other dictionaries).

Let's use data from `student_grades` again, but assume we just want the language grade so we can use the student's name as a key:

```
foreign_language_grades = {}
for student_name, subject, grade in student_grades:
    record = {
        'subject': subject,
        'grade': grade,
    }
    foreign_language_grades[student_name] = record

foreign_language_grades
```

```
{'Alice': {'subject': 'Spanish', 'grade': 'A'},
 'Bob': {'subject': 'French', 'grade': 'C'},
 'Carol': {'subject': 'Italian', 'grade': 'B+'},
 'Dave': {'subject': 'Italian', 'grade': 'A-'}}
```

Now we can refer to these by student name:

```
foreign_language_grades['Alice']
```
```
{'subject': 'Spanish', 'grade': 'A'}
```

And we can get the individual data that we care about:

```
foreign_language_grades['Alice']['grade']
```
```
'A'
```

A.7.4 Dictionary with Tuple Keys

It is occasionally useful to key dictionaries on more than one data. Dictionaries can use any immutable object as a key, which includes tuples. Continuing with our student grades example, we may want the keys to be the student name and subject:

```
course_grades = {}
for student_name, subject, grade in student_grades:
    course_grades[student_name, subject] = grade

course_grades
```
```
{('Alice', 'Spanish'): 'A',
 ('Bob', 'French'): 'C',
 ('Carol', 'Italian'): 'B+',
 ('Dave', 'Italian'): 'A-'}
```

Now we can represent all of a student's grades:

```
course_grades['Alice', 'Math'] = 'A'
course_grades['Alice', 'History'] = 'B'
course_grades
```
```
{('Alice', 'Spanish'): 'A',
 ('Bob', 'French'): 'C',
 ('Carol', 'Italian'): 'B+',
 ('Dave', 'Italian'): 'A-',
 ('Alice', 'Math'): 'A',
 ('Alice', 'History'): 'B'}
```

A.7.5 Another Dictionary of Dictionaries

For a particular student, we often want to get subject–grade pairs — a report card. We can create a dictionary with student names as keys and the values being dictionaries of subject–grade pairs. In this case we need to do a bit of checking — that step is commented below:

```
report_cards = {}
for student_name, subject, grade in student_grades:
    # If there is no report card for a student,
    # we need to create a blank one
    if student_name not in report_cards:
        report_cards[student_name] = {}
    report_cards[student_name][subject] = grade
report_cards
```
```
{'Alice': {'Spanish': 'A'},
 'Bob': {'French': 'C'},
 'Carol': {'Italian': 'B+'},
 'Dave': {'Italian': 'A-'}}
```

The advantage of this extra work is that we can now easily have multiple grades per student:

```
report_cards['Alice']['Math'] = 'A'
report_cards['Alice']['History'] = 'B'
report_cards
```
```
{'Alice': {'Spanish': 'A', 'Math': 'A', 'History': 'B'},
 'Bob': {'French': 'C'},
 'Carol': {'Italian': 'B+'},
 'Dave': {'Italian': 'A-'}}
```

And we can easily fetch a student's "report card":

```
report_cards['Alice']
```
```
{'Spanish': 'A', 'Math': 'A', 'History': 'B'}
```

NetLogo is a multi-agent programmable modeling environment. It is developed and maintained by the Center for Connected Learning and Computer-Based Modeling at Northwestern University (Wilensky, 1999) and is freely available for download as a desktop application[1] or for running on the Web[2] — we recommend the desktop version.

NetLogo comes with a large library of sample models, including several about networks. These pre-written models allow you to experiment without having to code an entire model. By playing with different initial configurations and parameters, you can observe how they affect the dynamics and outcomes of a model. In this way, you can gain a deeper understanding of the underlying rules and emergent network phenomena.

Once a model is loaded from the library (using the File menu in the application or the Search box on the Web version), you see three panels: the interface, information, and code tabs. The information tab introduces the model, explains how to use it, and suggests things to explore.

Let us briefly overview the key interface elements of a NetLogo model: buttons, switches, sliders, and monitors. These elements allow you to interact with the model. Buttons are used to set up, start, and stop the model. Sliders and switches alter model settings. Monitors and plots display data. To start the model, you first need to set it up with the `setup` button. Then you can run through the model one step at a time or loop through the iterations with the `go` button. A slider allows you to control the speed of execution. Switches and sliders give you access to a model's settings and parameters, so that you can explore different scenarios or hypotheses. The view lets you see what happens to the network being modeled. Plots and monitors show how key model statistics change over time. Plots have legends to interpret the meaning of the charts. You can export plot data to a spreadsheet.

Although it is possible to access and modify a model's source code (in the NetLogo programming language) via the code tab, and even to write your own models, here we focus on running a few library models that are most relevant to the material in this textbook.

B.1 PageRank

The PageRank model (Stonedahl and Wilensky, 2009) is illustrated in Figure B.1. PageRank is discussed in Section 4.3. The model demonstrates both an agent-based implementation of the random walk (`random-surfer`) model and the power (`diffusion`)

---

[1] ccl.northwestern.edu/netlogo/

[2] www.netlogoweb.org

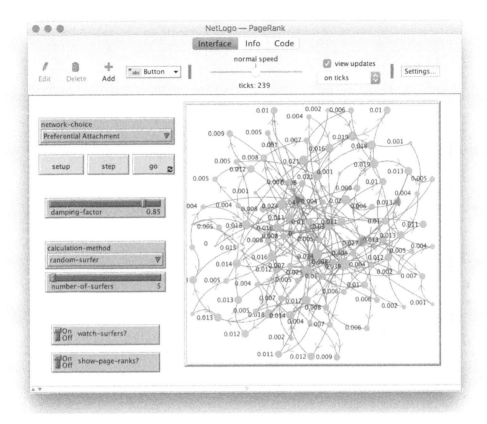

Screenshot of the NetLogo PageRank model. The model is licensed under CC BY-NC-SA 3.0 and reproduced with permission.

method for calculating PageRank. The random walk implementation has a parameter that allows you to specify the number of "surfers." You can see how these agents move or jump from page to page. Note the difference in speed between the two methods.

The model shows how PageRank is updated for each node at each iteration. The size of each node is roughly proportional to its PageRank. Choices of network include two simple example networks and a larger network with broader in-degree distribution, generated via preferential attachment. In one of the example networks, some of the nodes have no incoming links, yet they end up with non-zero PageRank. Play with the damping factor to see how it affects these values. Explore what happens when the damping factor is close to zero or one.

B.2 Giant Component

The Giant Component model (Wilensky, 2005a) is illustrated in Figure B.2. It demonstrates how quickly the giant component emerges in random networks when the average degree

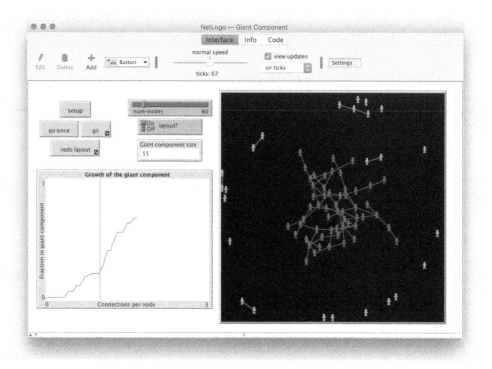

Screenshot of the NetLogo Giant Component model. The model is licensed under CC BY-NC-SA 3.0 and reproduced
with permission.

increases, as discussed in Section 5.1. Initially the link probability, average degree, and
density are all zero; there are no links and each node is a singleton. At each step, a link is
added between two random nodes that are not already connected by a link. As the model
runs, components are formed — small initially, then progressively larger as new links cause
distinct components to merge. The giant component is highlighted in red.

The only parameter of the model is the size of the network. A plot shows how the giant
component grows over time, as a function of the average degree. The vertical line on the
plot shows where the average degree equals one. Observe how the rate of growth of the
giant component increases after this critical point: the network undergoes a transition from
a fragmented phase with many small components to a mostly connected phase with a giant
component and a few remaining small components. Compare the behavior across multiple
runs with the same network size and with different sizes.

B.3 Small Worlds

The Small Worlds model (Wilensky, 2005c) is illustrated in Figure B.3. It implements
the small-world model discussed in Section 5.2, showing how to generate networks with
short average path length and high clustering coefficient. A parameter defines the number

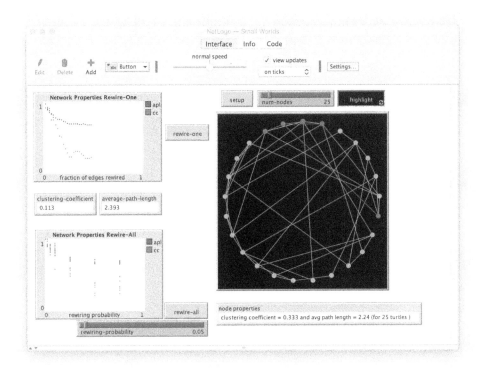

of nodes. After setup of the initial grid network, you can rewire one link at a time and observe how the average path length and clustering coefficient decrease as a function of the fraction of links rewired (top plot). A second mode is to set the rewiring probability parameter, and then rewire all links at once with that probability. The final average path length and clustering coefficient are charted versus the rewiring probability (bottom plot).

Play with different rewiring probabilities and observe the trends of the average path length and clustering coefficient values. Note that in certain ranges of the rewiring probability, the average path length decreases faster than the clustering coefficient. In fact, there is a range of values for which the (normalized) average path length is much smaller than the (normalized) clustering coefficient. Networks in that range are considered small worlds. Identify the approximate range and try to get a small world by rewiring one link at a time. Explore whether the trends depend on the number of nodes in the network.

B.4 Preferential Attachment

The Preferential Attachment model (Wilensky, 2005b) is illustrated in Figure B.4. It demonstrates how hubs emerge via preferential attachment, as discussed in Section 5.4. The model starts with two nodes connected by a link. At each step, a new node is added

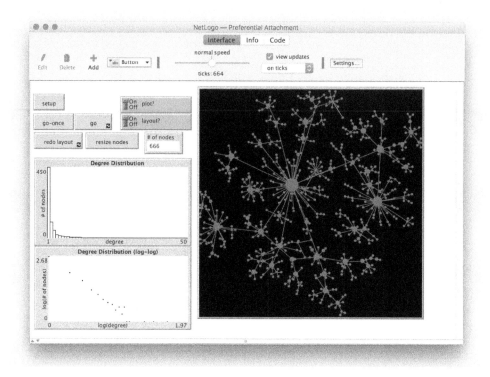

Screenshot of the NetLogo Preferential Attachment model. The model is licensed under CC BY-NC-SA 3.0 and
reproduced with permission.

and connected to a single existing node. The latter is selected randomly, but with some
bias: the selection probability is proportional to a node's degree. Note that, because only
one link is added for each new node, this implementation of the model yields trees (see
Section 2.4).

Use the `resize nodes` button, which makes node size proportional to degree, to
observe how hubs arise. Note whether older or newer nodes are more likely to become
the major hubs. You can study the degree distribution of the network by looking at the
plots. The top plot is a histogram of the node degrees. The bottom plot shows the same
distribution, but both axes are on a logarithmic scale. Let the model run for a while, then
describe the shape of the degree distribution in the log–log plot. Compare it with the distribution in Figure 5.8(c). Speed up the model (you can turn off the `layout` switch and
uncheck the "view updates" box) to allow a large network to form. Inspect how the degree
distribution gets broader as the network grows larger. Explain.

B.5 Virus on a Network

The Virus on a Network model (Stonedahl and Wilensky, 2008) combines the SIS
and SIR epidemic spreading models, discussed in Section 7.2.1. The parameter

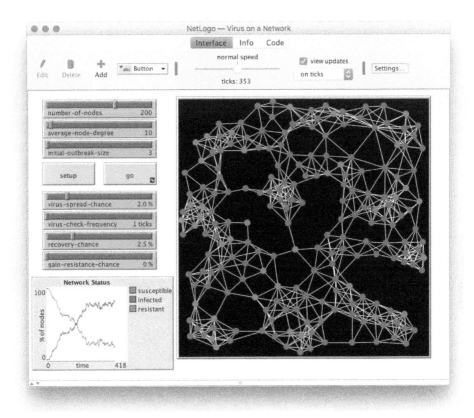

Screenshot of the NetLogo Virus on a Network model. The model is licensed under CC BY-NC-SA 3.0 and reproduced with permission.

`gain-resistance-chance` mixes the two models: when it is zero, the model corresponds to SIS (as illustrated in Figure B.5) and when it is one, the model corresponds to SIR. The model stops running once the virus has completely died out. The plot shows the numbers of nodes in the three states (S, I, R) over time. The links between resistant nodes and their neighbors are darkened, since they can no longer spread the virus. The network has homogeneous degree and geographic homophily: only nodes that are near each other (based on Euclidean distance) have a chance of being linked. Other parameters of the model include the average degree of the network, the infection rate (`virus-spread-chance`), and the recovery rate (`recovery-chance`).

Run the model with extreme values of `gain-resistance-chance` to reproduce SIS and SIR dynamics. In SIS, observe how the infection and recovery rates affect the balance between S and I populations. In SIR, explore the effects of the infection rate, recovery rate, and average degree on the maximum number of infected nodes. First vary each parameter while keeping the others constant, and note what happens. Then play with different combinations of the three parameters to reproduce the behaviors explained by the epidemic threshold in Eq. (7.5). Explain what conditions allow the epidemics to spread through most of the network. When the virus dies out without infecting the entire population, inspect the nodes that are spared. Even above the epidemic threshold, some clusters of nodes may never get infected. Describe the key structural characteristics of nodes, links, and communities that favor or hinder the epidemic spread.

B.6 Language Change

The Language Change model (Troutman and Wilensky, 2007) combines the voter and majority models of opinion dynamics (Section 7.3.1), as well as the fractional threshold model of social contagion (Section 7.1.1). The `update-algorithm` parameter lets you select among these models: the "individual" algorithm corresponds to the voter model, illustrated in Figure B.6. The "threshold" algorithm corresponds to the fractional threshold model, in which case you can set the threshold (`threshold-val`) parameter; when this is 0.5, the model is equivalent to the majority model. The model runs on a small preferential

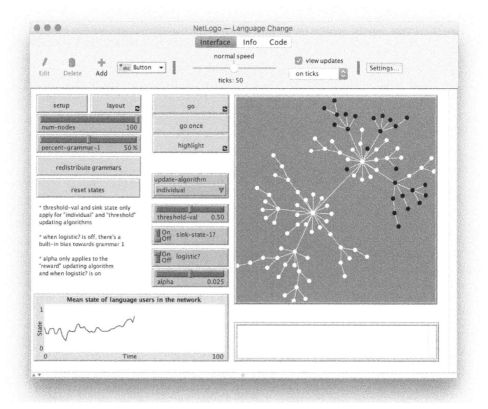

attachment tree (up to 100 nodes). The states, or opinions, are referred to as "grammars" zero (black) and one (white). Another key parameter is the initial fraction of nodes in state one (`percent-grammar-1`). Be sure to turn `sink-state-1` off, or else nodes cannot go back to state zero once they adopt state one.

In the threshold model, observe the role of the hubs as influencers. Then set the threshold to 0.5 and compare how the network converges to a stationary state in the voter and majority models. Note when the stationary state is one of consensus or polarization with coexisting states. Report on which model yields consensus more often. Explore whether other conditions, such as the number of nodes and initial configuration, affect the outcome. Study the probability of reaching white-state consensus as a function of the initial fraction of white nodes. Discuss whether this exit probability matches the behaviors in Figure 7.12 for the two models.

References

Achlioptas, D., Clauset, A., Kempe, D., and Moore, C. 2009. On the bias of traceroute sampling: Or, power-law degree distributions in regular graphs. *Journal of the ACM*, **56**(4), 21.

Adamic, L. A., Lukose, R. M., Puniyani, A. R., and Huberman, B. A. 2001. Search in power-law networks. *Physical Review E*, **64**(4), 046135.

Ahn, Y.-Y., Ahnert, S. E., Bagrow, J. P., and Barabási, A.-L. 2011. Flavor network and the principles of food pairing. *Scientific Reports*, **1**, 196.

Aiello, L., Barrat, A., Schifanella, R., Cattuto, C., Markines, B., and Menczer, F. 2012. Friendship prediction and homophily in social media. *ACM Transactions on the Web*, **6**(2), 9.

Albert, R., Jeong, H., and Barabási, A.-L. 1999. Internet: Diameter of the world-wide web. *Nature*, **401**(6749), 130.

Albert, R., Jeong, H., and Barabási, A.-L. 2000. Error and attack tolerance of complex networks. *Nature*, **406**(6794), 378–382.

Anderson, R. M., and May, R. M. 1992. *Infectious Diseases of Humans: Dynamics and Control*. Oxford University Press: Oxford.

Arenas, A., Duch, J., Fernández, A., and Gómez, S. 2007. Size reduction of complex networks preserving modularity. *New Journal of Physics*, **9**(6), 176.

Arenas, A., Fernández, A., and Gómez, S. 2008. Analysis of the structure of complex networks at different resolution levels. *New Journal of Physics*, **10**(5), 053039.

Backstrom, L., Boldi, P., Rosa, M., Ugander, J., and Vigna, S. 2012. Four degrees of separation. In *Proceedings of the 4th Annual ACM Web Science Conference* (WebSci '12), pp. 33–42.

Baeza-Yates, R., and Ribeiro-Neto, B. 2011. *Modern Information Retrieval: The Concepts and Technology Behind Search*, 2nd edn. ACM Press Books Addison-Wesley: New York.

Barabási, A.-L. 2003. *Linked: How Everything is Connected to Everything Else and What it Means for Business, Science, and Everyday Life*. Basic Books: New York.

Barabási, A.-L. 2016. *Network Science*. Cambridge University Press: Cambridge.

Barabási, A.-L., and Albert, R. 1999. Emergence of scaling in random networks. *Science*, **286**(5439), 509–512.

Barrat, A., Barthélemy, M., and Vespignani, A. 2008. *Dynamical Processes on Complex Networks*. Cambridge University Press: Cambridge.

Bastian, M., Heymann, S., Jacomy, M., *et al.* 2009. Gephi: An open source software for exploring and manipulating networks. In *Proceedings of the Third AAAI International Conference on Web and Social Media* (ICWSM), pp. 361–362.

Batagelj, V., Mrvar, A., and Zaveršnik, M. 1999. Partitioning approach to visualization of large graphs. In *International Symposium on Graph Drawing*, pp. 90–97.

Baur, M., Brandes, U., Gaertler, M., and Wagner, D. 2004. Drawing the AS graph in 2.5 dimensions. In *International Symposium on Graph Drawing*, pp. 43–48.

Bavelas, A. 1950. Communication patterns in task-oriented groups. *Journal of the Acoustical Society of America*, **22**(6), 725–730.

Beiró, M. G., Alvarez-Hamelin, J. I., and Busch, J. R. 2008. A low complexity visualization tool that helps to perform complex systems analysis. *New Journal of Physics*, **10**(12), 125003.

Bellman, R. 1958. On a routing problem. *Quarterly of Applied Mathematics*, **16**, 87–90.

Berners-Lee, T., and Fischetti, M. 2000. *Weaving the Web: The Original Design and Ultimate Destiny of the World Wide Web by its Inventor*. HarperCollins: New York.

Bhan, A., Galas, D. J., and Dewey, T. G. 2002. A duplication growth model of gene expression networks. *Bioinformatics*, **18**(11), 1486–1493.

Bianconi, G., and Barabási, A.-L. 2001. Bose–Einstein condensation in complex networks. *Physical Review Letters*, **86**(24), 5632–5635.

Bichot, C.-E., and Siarry, P. 2013. *Graph Partitioning*. Wiley: Hoboken, NJ.

Blondel, V. D., Guillaume, J.-L., Lambiotte, R., and Lefebvre, E. 2008. Fast unfolding of communities in large networks. *Journal of Statistical Mechanics*, **P10008**.

Boccaletti, S., Bianconi, G., Criado, R., Del Genio, C. I., Gómez-Gardenes, J., Romance, M., *et al.* 2014. The structure and dynamics of multilayer networks. *Physics Reports*, **544**(1), 1–122.

Bollobás, B. 2012. *Graph Theory: An Introductory Course*. Springer: New York.

Brandes, U. 2001. A faster algorithm for betweenness centrality. *Journal of Mathematical Sociology*, **25**(2), 163–177.

Brin, S., and Page, L. 1998. The anatomy of a large-scale hypertextual web search engine. *Computer Networks and ISDN Systems*, **30**(1–7), 107–117.

Broder, A., Kumar, R., Maghoul, F., Raghavan, P., Rajagopalan, S., Stata, R., *et al.* 2000. Graph structure in the web. *Computer Networks*, **33**(1–6), 309–320.

Caldarelli, G. 2007. *Scale-Free Networks*. Oxford University Press: Oxford.

Caldarelli, G., and Chessa, A. 2016. *Data Science and Complex Networks: Real Case Studies with Python*. Oxford University Press: Oxford.

Castellano, C., Fortunato, S., and Loreto, V. 2009. Statistical physics of social dynamics. *Reviews of Modern Physics*, **81**(2), 591–646.

Centola, D., and Macy, M. 2007. Complex contagions and the weakness of long ties. *American Journal of Sociology*, **113**(3), 702–734.

Cha, M., Haddadi, H., Benevenuto, F., and Gummadi, K. P. 2010. Measuring user influence in Twitter: The million follower fallacy. In *Proceedings of the 4th International AAAI Conference on Weblogs and Social Media* (ICWSM), pp. 10–17.

Christakis, N. A., and Fowler, J. H. 2010. Social network sensors for early detection of contagious outbreaks. *PloS ONE*, **5**(9), e12948.

Clauset, A., Newman, M. E. J., and Moore, C. 2004. Finding community structure in very large networks. *Physical Review E*, **70**(6), 066111.

Clifford, P., and Sudbury, A. 1973. A model for spatial conflict. *Biometrika*, **60**(3), 581–588.

Cohen, R., and Havlin, S. 2003. Scale-free networks are ultrasmall. *Physical Review Letters*, **90**(5), 058701.

Cohen, R., and Havlin, S. 2010. *Complex Networks: Structure, Robustness and Function*. Cambridge University Press: Cambridge.

Cohen, R., Erez, K., Ben-Avraham, D., and Havlin, S. 2000. Resilience of the Internet to random breakdowns. *Physical Review Letters*, **85**(21), 4626–4628.

Cohen, R., Erez, K., Ben-Avraham, D., and Havlin, S. 2001. Breakdown of the Internet under intentional attack. *Physical Review Letters*, **86**(16), 3682–3685.

Cohen, R., Havlin, S., and Ben-Avraham, D. 2002. Structural properties of scale-free networks. *Handbook of Graphs and Networks: From the Genome to the Internet*. Wiley: Weinheim.

Cohen, R., Havlin, S., and Ben-Avraham, D. 2003. Efficient immunization strategies for computer networks and populations. *Physical Review Letters*, **91**(24), 247901.

Condon, A., and Karp, R. M. 2001. Algorithms for graph partitioning on the planted partition model. *Random Structures and Algorithms*, **18**, 116–140.

Conover, M., Gonçalves, B., Ratkiewicz, J., Flammini, A., and Menczer, F. 2011a. Predicting the political alignment of Twitter users. In *Proceedings of the 3rd IEEE Conference on Social Computing* (SocialCom), pp. 192–199.

Conover, M., Ratkiewicz, J., Francisco, M., Gonçalves, B., Flammini, A., and Menczer, F. 2011b. Political polarization on Twitter. In *Proceedings of the 5th International AAAI Conference on Weblogs and Social Media* (ICWSM), pp. 89–96.

Daley, D. J., and Kendall, D. G. 1964. Epidemics and rumours. *Nature*, **204**(4963), 1118.

Davison, B. D. 2000. Topical locality in the web. In *Proceedings of the 23rd Annual International ACM Conference on Research and Development in Information Retrieval* (SIGIR), pp. 272–279.

Dawkins, R. 2016. *The Selfish Gene: 40th Anniversary Edition*, 4th edn. Oxford University Press: Oxford.

Deffuant, G., Neau, D., Amblard, F., and Weisbuch, G. 2000. Mixing beliefs among interacting agents. *Advances in Complex Systems*, **3**(01n04), 87–98.

Di Battista, G., Eades, P., Tamassia, R., and Tollis, I. G. 1998. *Graph Drawing: Algorithms for the Visualization of Graphs*. Prentice Hall: Upper Saddle River, NJ.

Dijkstra, E. W. 1959. A note on two problems in connexion with graphs. *Numerische Mathematik*, **1**(1), 269–271.

Dodds, P. S., Muhamad, R., and Watts, D. J. 2003. An experimental study of search in global social networks. *Science*, **301**(5634), 827–829.

Dorogovtsev, S. N., and Mendes, J. F. F. 2013. *Evolution of Networks: From Biological Nets to the Internet and WWW*. Oxford University Press: Oxford.

Dorogovtsev, S. N., Mendes, J. F. F., and Samukhin, A. N. 2000. Structure of growing networks with preferential linking. *Physical Review Letters*, **85**(21), 4633–4636.

Dunbar, R. I. M. 1992. Neocortex size as a constraint on group size in primates. *Journal of Human Evolution*, **22**(6), 469–493.

Dunne, J. A., Williams, R. J., and Martinez, N. D. 2002. Food-web structure and network theory: The role of connectance and size. *Proceedings of the National Academy of Sciences of the USA*, **99**(20), 12917–12922.

Eades, P. 1984. A heuristic for graph drawing. *Congressus Numerantium*, **42**, 149–160.

Easley, D., and Kleinberg, J. 2010. *Networks, Crowds, and Markets: Reasoning About a Highly Connected World*. Cambridge University Press: Cambridge.

Erdös, P., and Rényi, A. 1959. On random graphs. I. *Publicationes Mathematical Debrecen*, **6**, 290–297.

Feld, S. L. 1991. Why your friends have more friends than you do. *American Journal of Sociology*, **96**(6), 1464–1477.

Ferrara, E., Varol, O., Davis, C., Menczer, F., and Flammini, A. 2016. The rise of social bots. *Communications of the ACM*, **59**(7), 96–104.

Festinger, L. 1954. A theory of social comparison processes. *Human Relations*, **7**(2), 117–140.

Fienberg, S. E., and Wasserman, S. 1981. Categorical data analysis of single sociometric relations. *Sociological Methodology*, **12**, 156–192.

Ford Jr., L. R. 1956. *Network Flow Theory*. Technical Report Paper P-923. RAND Corporation.

Fortunato, S. 2010. Community detection in graphs. *Physics Reports*, **486**(3–5), 75–174.

Fortunato, S., and Barthélemy, M. 2007. Resolution limit in community detection. *Proceedings of the National Academy of Sciences of the USA*, **104**(1), 36–41.

Fortunato, S., and Hric, D. 2016. Community detection in networks: A user guide. *Physics Reports*, **659**, 1–44.

Fortunato, S., Flammini, A., and Menczer, F. 2006. Scale-free network growth by ranking. *Physical Review Letters*, **96**(21), 218701.

Fortunato, S., Boguñá, M., Flammini, A., and Menczer, F. 2007. On local estimations of PageRank: A mean field approach. *Internet Mathematics*, **4**(2–3), 245–266.

Fred, A. L. N., and Jain, A. K. 2003. Robust data clustering. In *Proceedings of the 2003 IEEE Computer Society Conference on Computer Vision and Pattern Recognition*, pp. 128–136.

Freedman, D., Pisani, R., and Purves, R. 2007. *Statistics*. W. W. Norton & Co.: New York.

Freeman, L. C. 1977. A set of measures of centrality based on betweenness. *Sociometry*, 40(1), 35–41.

Fruchterman, T. M. J., and Reingold, E. M. 1991. Graph drawing by force-directed placement. *Software: Practice and Experience*, **21**(11), 1129–1164.

Galam, S. 2002. Minority opinion spreading in random geometry. *The European Physical Journal B: Condensed Matter and Complex Systems*, **25**(4), 403–406.

Gao, J., Buldyrev, S. V., Stanley, H. E., and Havlin, S. 2012. Networks formed from interdependent networks. *Nature Physics*, **8**(1), 40–48.

Gil, S., and Zanette, D. H. 2006. Coevolution of agents and networks: Opinion spreading and community disconnection. *Physics Letters A*, **356**(2), 89–94.

Gilbert, E. N. 1959. Random graphs. *Annals of Mathematical Statistics*, **30**(4), 1141–1144.

Girvan, M., and Newman, M. E. J. 2002. Community structure in social and biological networks. *Proceedings of the National Academy of Sciences of the USA*, **99**(12), 7821–7826.

Glauber, R. J. 1963. Time-dependent statistics of the Ising model. *Journal of Mathematical Physics*, **4**(2), 294–307.

Gleich, D. F. 2015. PageRank beyond the Web. *SIAM Review*, **57**(3), 321–363.

Goel, S., Anderson, A., Hofman, J., and Watts, D. J. 2015. The structural virality of online diffusion. *Management Science*, **62**(1), 180–196.

Goldenberg, J., Libai, B., and Muller, E. 2001. Talk of the network: A complex systems look at the underlying process of word-of-mouth. *Marketing Letters*, **12**(3), 211–223.

Granovetter, M. 1973. The strength of weak ties. *American Journal of Sociology*, **78**(6), 1360–1380.

Granovetter, M. 1978. Threshold models of collective behavior. *American Journal of Sociology*, **83**(6), 1420–1443.

Guimerà, R., Sales-Pardo, M., and Amaral, L. A. 2004. Modularity from fluctuations in random graphs and complex networks. *Physical Review E*, **70**(2), 025101(R).

Holland, P. W., and Leinhardt, S. 1971. Transitivity in structural models of small groups. *Comparative Group Studies*, **2**(2), 107–124.

Holland, P. W., and Leinhardt, S. 1981. An exponential family of probability distributions for directed graphs. *Journal of the American Statistical Association*, **76**(373), 33–50.

Holland, P., Laskey, K. B., and Leinhardt, S. 1983. Stochastic blockmodels: First steps. *Social Networks*, **5**(2), 109–137.

Holme, P., and Newman, M. E. J. 2006. Nonequilibrium phase transition in the coevolution of networks and opinions. *Physical Review E*, **74**(5), 056108.

Holme, P., and Saramäki, J. 2012. Temporal networks. *Physics Reports*, **519**(3), 97–125.

Hric, D., Darst, R. K., and Fortunato, S. 2014. Community detection in networks: Structural communities versus ground truth. *Physical Review E*, **90**(6), 062805.

Hu, Y., Chen, H., Zhang, P., Li, M., Di, Z., and Fan, Y. 2008. Comparative definition of community and corresponding identifying algorithm. *Physical Review E*, **78**(2), 026121.

Jacomy, M., Venturini, T., Heymann, S., and Bastian, M. 2014. ForceAtlas2, a continuous graph layout algorithm for handy network visualization designed for the Gephi software. *PloS ONE*, **9**(6), e98679.

Jagatic, T. N., Johnson, N. A., Jakobsson, M., and Menczer, F. 2007. Social phishing. *Communications of the ACM*, **50**(10), 94–100.

Jain, A. K., Murty, M. N., and Flynn, P. J. 1999. Data clustering: A review. *ACM Computing Surveys*, **31**(3), 264–323.

Jeong, H., Mason, S. P., Barabási, A.-L., and Oltvai, Z. N. 2001. Lethality and centrality in protein networks. *Nature*, **411**(6833), 41–42.

Jernigan, C., and Mistree, B. F. T. 2009. Gaydar: Facebook friendships expose sexual orientation. *First Monday*, **14**(10).

Kamada, T., and Kawai, S. 1989. An algorithm for drawing general undirected graphs. *Information Processing Letters*, **31**(1), 7–15.

Karrer, B., and Newman, M. E. J. 2011. Stochastic blockmodels and community structure in networks. *Physical Review E*, **83**(1), 016107.

Kempe, D., Kleinberg, J., and Tardos, É. 2003. Maximizing the spread of influence through a social network. In *Proceedings of the Ninth ACM SIGKDD International Conference on Knowledge Discovery and Data Mining*, pp. 137–146.

Kernighan, B. W., and Lin, S. 1970. An efficient heuristic procedure for partitioning graphs. *Bell System Technical Journal*, **49**(2), 291–307.

Kitsak, M., Gallos, L. K., Havlin, S., Liljeros, F., Muchnik, L., Stanley, H. E., and Makse, H. A. 2010. Identification of influential spreaders in complex networks. *Nature Physics*, **6**(11), 888–893.

Kivelä, M., Arenas, A., Barthélemy, M., Gleeson, J. P., Moreno, Y., and Porter, M. A. 2014. Multilayer networks. *Journal of Complex Networks*, **2**(3), 203–271.

Kleinberg, J. M. 1999. Authoritative sources in a hyperlinked environment. *Journal of the ACM*, **46**(5), 604–632.

Kleinberg, J. M. 2000. Navigation in a small world. *Nature*, **406**(6798), 845.

Kleinberg, J. M. 2002. Small-world phenomena and the dynamics of information. In *Advances in Neural Information Processing Systems: Proceedings of the First 12 Conferences*, pp. 431–438.

Kleinberg, J. M, Kumar, R., Raghavan, P., Rajagopalan, S., and Tomkins, A. S. 1999. The web as a graph: Measurements, models, and methods. In *Computing and Combinatorics: Proceedings of the 5th Annual International Conference*, pp. 1–17.

Krapivsky, P. L., and Redner, S. 2001. Organization of growing random networks. *Physical Review E*, **63**(6), 066123.

Krapivsky, P. L., and Redner, S. 2003. Dynamics of majority rule in two-state interacting spin systems. *Physical Review Letters*, **90**(23), 238701.

Krapivsky, P. L., Redner, S., and Leyvraz, F. 2000. Connectivity of growing random networks. *Physical Review Letters*, **85**(21), 4629–4632.

Lancichinetti, A., and Fortunato, S. 2009. Community detection algorithms: A comparative analysis. *Physical Review E*, **80**(5), 056117.

Lancichinetti, A., Fortunato, S., and Radicchi, F. 2008. Benchmark graphs for testing community detection algorithms. *Physical Review E*, **78**(4), 046110.

Latora, V., Nicosia, V., and Russo, G. 2017. *Complex Networks: Principles, Methods and Applications*. Cambridge University Press: Cambridge.

Lazarsfeld, P. F., Merton, R. K., *et al.* 1954. Friendship as a social process: A substantive and methodological analysis. *Freedom and Control in Modern Society*, **18**(1), 18–66.

Lazer, D. M. J., Baum, M. A., Benkler, Y., Berinsky, A. J., Greenhill, K. M., Menczer, F., *et al.* 2018. The science of fake news. *Science*, **359**(6380), 1094–1096.

Liljeros, F., Edling, C. R., Amaral, L. A. N., Stanley, H. E., and Åberg, Y. 2001. The web of human sexual contacts. *Nature*, **411**, 907–908.

Liu, B. 2011. *Web Data Mining: Exploring Hyperlinks, Contents, and Usage Data*, 2nd edn. Springer: New York.

Luccio, F., and Sami, M. 1969. On the decomposition of networks into minimally interconnected networks. *IEEE Transactions on Circuit Theory*, **16**(2), 184–188.

Luce, R. D., and Perry, A. D. 1949. A method of matrix analysis of group structure. *Psychometrika*, **14**(2), 95–116.

Manning, C. D., Raghavan, P., and Schütze, H. 2008. *Introduction to Information Retrieval.* Cambridge University Press: Cambridge.

Marchiori, M. 1997. The quest for correct information on the web: Hyper search engines. *Computer Networks and ISDN Systems*, **29**(8–13), 1225–1235.

McPherson, M., Smith-Lovin, L., and Cook, J. M. 2001. Birds of a feather: Homophily in social networks. *Annual Review of Sociology*, **27**(1), 415–444.

Meilă, M. 2007. Comparing clusterings—an information based distance. *Journal of Multivariate Analysis*, **98**(5), 873–895.

Meiss, M., Menczer, F., Fortunato, S., Flammini, A., and Vespignani, A. 2008. Ranking web sites with real user traffic. In *Proceedings of the 1st ACM International Conference on Web Search and Data Mining* (WSDM), pp. 65–75.

Meiss, M., Gonçalves, B., Ramasco, J., Flammini, A., and Menczer, F. 2010. Modeling traffic on the web graph. In *Proceedings of the 7th Workshop on Algorithms and Models for the Web Graph* (WAW), pp. 50–61.

Melián, C. J., and Bascompte, J. 2004. Food web cohesion. *Ecology*, **85**(2), 352–358.

Menczer, F. 2002. Growing and navigating the small world web by local content. *Proceedings of the National Academy of Sciences of the USA*, **99**(22), 14014–14019.

Menczer, F. 2004. Lexical and semantic clustering by web links. *Journal of the American Society for Information Science and Technology*, **55**(14), 1261–1269.

Meusel, R., Vigna, S., Lehmberg, O., and Bizer, C. 2015. The graph structure in the web — analyzed on different aggregation levels. *Journal of Web Science*, **1**(1), 33–47.

Mobilia, M., Petersen, A., and Redner, S. 2007. On the role of zealotry in the voter model. *Journal of Statistical Mechanics: Theory and Experiment*, P08029.

Molloy, M., and Reed, B. 1995. A critical point for random graphs with a given degree sequence. *Random Structures and Algorithms*, **6**(2–3), 161–179.

Moore, E. F. 1959. The shortest path through a maze. In *Proceedings of the International Symposium on Switching Theory 1957, Part II*, pp. 285–292.

Moreno, J. L., and Jennings, H. H. 1934. *Who Shall Survive?* Nervous and Mental Disease Publishing Co.: New York.

Newman, M. E. J. 2001. The structure of scientific collaboration networks. *Proceedings of the National Academy of Sciences of the USA*, **98**(2), 404–409.

Newman, M. E. J. 2002. Assortative mixing in networks. *Physical Review Letters*, **89**(20), 208701.

Newman, M. E. J. 2004a. Fast algorithm for detecting community structure in networks. *Physical Review E*, **69**(6), 066133.

Newman, M. E. J. 2004b. Analysis of weighted networks. *Physical Review E*, **70**(5), 056131.

Newman, M. 2018. *Networks*, 2nd edn. Oxford University Press: Oxford.

Newman, M. E. J., and Girvan, M. 2004. Finding and evaluating community structure in networks. *Physical Review E*, **69**(2), 026113.

Pariser, E. 2011. *The Filter Bubble: What the Internet is Hiding From You.* Penguin: Harmondsworth.

Pastor-Satorras, R., and Vespignani, A. 2001. Epidemic spreading in scale-free networks. *Physical Review Letters*, **86**(14), 3200–3203.

Pastor-Satorras, R., and Vespignani, A. 2007. *Evolution and Structure of the Internet: A Statistical Physics Approach*. Cambridge University Press: Cambridge.

Pastor-Satorras, R., Vázquez, A., and Vespignani, A. 2001. Dynamical and correlation properties of the Internet. *Physical Review Letters*, **87**(25), 258701.

Pastor-Satorras, R., Castellano, C., Van Mieghem, P., and Vespignani, A. 2015. Epidemic processes in complex networks. *Reviews of Modern Physics*, **87**(3), 925–979.

Peixoto, T. P. 2012. Entropy of stochastic blockmodel ensembles. *Physical Review E*, **85**(5), 056122.

Peixoto, T. P. 2014. Hierarchical block structures and high-resolution model selection in large networks. *Physical Review X*, **4**(1), 011047.

Porter, M. A., Onnela, J.-P., and Mucha, P. J. 2009. Communities in networks. *Notices of the American Mathematical Society*, **56**(9), 1082–1097.

Price, D. D. 1976. A general theory of bibliometric and other cumulative advantage processes. *Journal of the American Society of Information Science*, **27**(5), 292–306.

Radicchi, F. 2015. Percolation in real interdependent networks. *Nature Physics*, **11**(7), 597–602.

Radicchi, F., Castellano, C., Cecconi, F., Loreto, V., and Parisi, D. 2004. Defining and identifying communities in networks. *Proceedings of the National Academy of Sciences of the USA*, **101**(9), 2658–2663.

Raghavan, U. N., Albert, R., and Kumara, S. 2007. Near linear time algorithm to detect community structures in large-scale networks. *Physical Review E*, **76**(3), 036106.

Ratkiewicz, J., Conover, M., Meiss, M., Gonçalves, B., Flammini, A., and Menczer, F. 2011. Detecting and tracking political abuse in social media. In *Proceedings of the 5th International AAAI Conference on Weblogs and Social Media* (ICWSM), pp. 297–304.

Reichardt, J., and Bornholdt, S. 2006. Statistical mechanics of community detection. *Physical Review E*, **74**(1), 016110.

Reis, S. D. S., Hu, Y., Babino, A., Andrade Jr., J. S., Canals, S., Sigman, M., and Makse, H. A. 2014. Avoiding catastrophic failure in correlated networks of networks. *Nature Physics*, **10**(10), 762–767.

Rossi, R. A., and Ahmed, N. K. 2015. The network data repository with interactive graph analytics and visualization. In *Proceedings of the 29th AAAI Conference on Artificial Intelligence*, pp. 4292–4293.

Rossi, R. A., Fahmy, S., and Talukder, N. 2013. A multi-level approach for evaluating Internet topology generators. In *Proceedings of the IFIP Networking Conference*, pp. 1–9.

Seeley, J. R. 1949. The net of reciprocal influence: A problem in treating sociometric data. *Canadian Journal of Experimental Psychology*, **3**(4), 234–240.

Serrano, M, Maguitman, A., Boguñá, M., Fortunato, S., and Vespignani, A. 2007. Decoding the structure of the WWW: A comparative analysis of Web crawls. *ACM Transactions on the Web*, **1**(2), 10.

Serrano, M. Á., Boguñá, M., and Vespignani, A. 2009. Extracting the multiscale backbone of complex weighted networks. *Proceedings of the National Academy of Sciences of the USA*, **106**(16), 6483–6488.

Shao, C., Hui, P.-M., Wang, L., Jiang, X., Flammini, A., Menczer, F., and Ciampaglia, G. L. 2018a. Anatomy of an online misinformation network. *PLoS ONE*, **13**(4), e0196087.

Shao, C., Ciampaglia, G. L., Varol, O., Yang, K., Flammini, A., and Menczer, F. 2018b. The spread of low-credibility content by social bots. *Nature Communications*, **9**, 4787.

Shimbel, A. 1955. Structure in communication nets. In *Proceedings of the Symposium on Information Networks*, pp. 199–203.

Solé, R. V., Pastor-Satorras, R., Smith, E., and Kepler, T. B. 2002. A model of large-scale proteome evolution. *Advances in Complex Systems*, **5**(01), 43–54.

Solomonoff, R., and Rapoport, A. 1951. Connectivity of random nets. *The Bulletin of Mathematical Biophysics*, **13**(2), 107–117.

Sporns, O. 2012. *Discovering the Human Connectome*. MIT Press: Boston, MA.

Spring, N., Mahajan, R., and Wetherall, D. 2002. Measuring ISP topologies with Rocketfuel. In *ACM SIGCOMM Computer Communication Review*, pp. 133–145.

Stehlé, J., Voirin, N., Barrat, A., Cattuto, C., Isella, L., Pinton, J.-F., *et al.* 2011. High-resolution measurements of face-to-face contact patterns in a primary school. *PLoS ONE*, **6**(8), e23176.

Stonedahl, F., and Wilensky, U. 2008. *NetLogo Virus on a Network Model*. Center for Connected Learning and Computer-Based Modeling, Northwestern University, Evanston, IL. http://ccl.northwestern.edu/netlogo/models/VirusonaNetwork.

Stonedahl, F., and Wilensky, U. 2009. *NetLogo PageRank Model*. Center for Connected Learning and Computer-Based Modeling, Northwestern University, Evanston, IL. http://ccl.northwestern.edu/netlogo/models/PageRank.

Sunstein, C. R. 2001. *Echo Chambers: Bush v. Gore, Impeachment, and Beyond*. Princeton University Press: Princeton, NJ.

Travers, J., and Milgram, S. 1969. An experimental study of the small world problem. *Sociometry*, **32**(4), 425–443.

Troutman, C., and Wilensky, U. 2007. *NetLogo Language Change Model*. Center for Connected Learning and Computer-Based Modeling, Northwestern University, Evanston, IL. http://ccl.northwestern.edu/netlogo/models/LanguageChange.

Ulanowicz, R. E., and DeAngelis, D. L. 1998. Network analysis of trophic dynamics in South Florida ecosystems. *FY97: The Florida Bay Ecosystem*, 20688–20038.

Vázquez, A. 2003. Growing network with local rules: Preferential attachment, clustering hierarchy, and degree correlations. *Physical Review E*, **67**(5), 056104.

Vázquez, A., Flammini, A., Maritan, A., and Vespignani, A. 2003a. Modeling of protein interaction networks. *Complexus*, **1**(1), 38–44.

Vázquez, F., Krapivsky, P. L., and Redner, S. 2003b. Constrained opinion dynamics: Freezing and slow evolution. *Journal of Physics A: Mathematical and General*, **36**(3), L61–L68.

Vosoughi, S., Roy, D., and Aral, S. 2018. The spread of true and false news online. *Science*, **359**(6380), 1146–1151.

Wagner, A. 1994. Evolution of gene networks by gene duplications: A mathematical model and its implications on genome organization. *Proceedings of the National Academy of Sciences of the USA*, **91**(10), 4387–4391.

Wasserman, S., and Faust, K. 1994. *Social Network Analysis: Methods and Applications*. Cambridge University Press: Cambridge.

Watts, D. J. 2002. A simple model of global cascades on random networks. *Proceedings of the National Academy of Sciences of the USA*, **99**(9), 5766–5771.

Watts, D. J. 2004. *Six Degrees: The Science of a Connected Age*. W. W. Norton & Co.: New York.

Watts, D. J., and Strogatz, S. H. 1998. Collective dynamics of 'small-world' networks. *Nature*, **393**(6684), 440–442.

Watts, D. J., Dodds, P. S., and Newman, M. E. J. 2002. Identity and search in social networks. *Science*, **296**(5571), 1302–1305.

Weng, L., Ratkiewicz, J., Perra, N., Gonçalves, B., Castillo, C., Bonchi, F., *et al.* 2013a. The role of information diffusion in the evolution of social networks. In *Proceedings of the 19th ACM SIGKDD Conference on Knowledge Discovery and Data Mining* (KDD), pp. 356–364.

Weng, L., Menczer, F., and Ahn, Y.-Y. 2013b. Virality prediction and community structure in social networks. *Scientific Reports*, **3**, 2522.

White, J. G., Southgate, E., Thomson, J. N., and Brenner, S. 1986. The structure of the nervous system of the nematode *Caenorhabditis elegans*. *Philosophical Transactions of the Royal Society of London Series B, Biological Science*, **314**(1165), 1–340.

Wilensky, U. 1999. *NetLogo*. Center for Connected Learning and Computer-Based Modeling, Northwestern University, Evanston, IL. http://ccl.northwestern.edu/netlogo/.

Wilensky, U. 2005a. *NetLogo Giant Component Model*. Center for Connected Learning and Computer-Based Modeling, Northwestern University, Evanston, IL. http://ccl.northwestern.edu/netlogo/models/GiantComponent.

Wilensky, U. 2005b. *NetLogo Preferential Attachment Model*. Center for Connected Learning and Computer-Based Modeling, Northwestern University, Evanston, IL. http://ccl.northwestern.edu/netlogo/models/PreferentialAttachment.

Wilensky, U. 2005c. *NetLogo Small Worlds Model*. Center for Connected Learning and Computer-Based Modeling, Northwestern University, Evanston, IL. http://ccl.northwestern.edu/netlogo/models/SmallWorlds.

Xu, R, and Wunsch, D. 2008. *Clustering*. Wiley: Piscataway, NJ.

Yang, J., and Leskovec, J. 2012. Defining and evaluating network communities based on ground-truth. In *Proceedings of the ACM SIGKDD Workshop on Mining Data Semantics* (MDS '12), pp. 3:1–3:8.

Zachary, W. W. 1977. An information flow model for conflict and fission in small groups. *Journal of Anthropological Research*, **33**(4), 452–473.

Index